a life science unit for high-ability learners in kindergarten and first grade

# Survive and Thrive

a life science unit for high-ability learners in kindergarten and first grade

# Survive and Thrive

Project Clarion Primary Science Units
Funded by the Jacob K. Javits Program, United States Department of Education

The College of William and Mary
School of Education
Center for Gifted Education
P.O. Box 8795
Williamsburg, VA 23187-8795

Co-Principal Investigators: Bruce A. Bracken & Joyce VanTassel-Baska
Project Directors: Lori C. Bland, Tamra Stambaugh, & Valerie Gregory
Unit Developers: Elizabeth B. Sutton & J. Denise Drain
Unit Editors: Steve V. Coxon & Cristina Reintjes

Routledge
Taylor & Francis Group

NEW YORK AND LONDON

First published in 2010 by Prufrock Press Inc.

Published in 2021 by Routledge
605 Third Avenue, New York, NY 10017
2 Park Square, Milton Park, Abingdon, Oxon OX14 4RN

*Routledge is an imprint of the Taylor & Francis Group, an informa business*

Copyright ©2010 Center for Gifted Education, The College of William and Mary

Production Design by Marjorie Parker

ISBN: 9781032142265 (hbk)
ISBN: 9781593633936 (pbk)

DOI: 10.4324/9781003238386

# Contents

## Part I: Unit Overview

Introduction to the Clarion Units............................................. 3

Teacher's Guide to Content................................................... 10

Unit Glossary............................................................... 12

Teaching Resources.......................................................... 13

## Part II: Lesson Plans

Lesson Plans ............................................................... 15

Overview of Lessons......................................................... 16

Preassessment .............................................................. 24

Lesson 1: What Is a Scientist?.............................................. 41

Lesson 2: What Is Change? .................................................. 45

Lesson 3: What Scientists Do—Observe, Question, Learn More.............. 50

Lesson 4: What Scientists Do—Experiment, Create Meaning, Tell Others..... 54

Lesson 5: What Is a Life Cycle?............................................. 61

Lesson 6: What Is the Life Cycle of a Mealworm? ............................ 67

Lesson 7: What Are the Requirements for Life? .............................. 81

Lesson 8: What Do Plants Need to Thrive? ................................... 85

Lesson 9: How Do Animals Look Different? ................................... 90

Lesson 10: What Is an Appendage?........................................... 95

Lesson 11: How Can We Classify Animals? ................................... 101

Lesson 12: What Have We Learned About Animals and Plants? ............ 106

Postassessment ............................................................. 110

Appendix A: Concept Paper on Change ...................................... 123

Appendix B: Teaching Models............................................... 127

Appendix C: Basic Concepts in Early Childhood ........................... 134

Appendix D: Materials List ................................................. 142

References ................................................................. 145

Next Generation Science Standards Alignment .............................. 147

# Part I: Unit Overview

# Introduction to the Clarion Units

The Project Clarion Science Units for Primary Grades introduce young students to science concepts, science reasoning, and scientific investigation processes. Macroconcepts, such as systems or change, help students connect understanding of science content and processes. The units use a hands-on, constructivist approach that allows children to build their knowledge base and their skills as they explore science topics through play and planned investigations. Students are engaged in creative and critical thinking, problem finding and solving, process skill development, and communication opportunities. Conceptual understanding is reinforced as units strengthen basic language and mathematical concepts, including quantity, direction, position, comparison, colors, letter identification, numbers, counting, size, social awareness, texture, material, shape, time, and sequence.

## Introduction to the *Survive and Thrive* Unit

*Survive and Thrive*, a kindergarten and first-grade life science unit, engages students in a study of animals, their characteristics, and their natural environments. Students learn how to distinguish features and life needs of animals and observe animals in their habitats via webcams. Students learn to classify animals according to whether they are tame or wild and live on land or in water. Students raise mealworms in the classroom and observe their life cycle. Focusing on the macroconcept of change, *Survive and Thrive* deepens students' understanding of the scientific concepts in the unit.

## Curriculum Framework

The curriculum framework (see Table 1) developed for the Project Clarion science units is based on the Integrated Curriculum Model (ICM), which posits the relatively equal importance of teaching to high-level content, higher order processes and resultant products, and important concepts and issues. The model represents a research-based set of differentiated curricular and instructional approaches found appropriate for high-ability learners (VanTassel-Baska, 1986; VanTassel-Baska & Little, 2003). The framework serves several important functions:

1. The curriculum framework provides scaffolding for the central concept of change, the scientific research process, and the content of the units.
2. The curriculum framework also provides representative statements of advanced, complex, and sophisticated learner outcomes. It demonstrates how a single set of outcomes for all can be translated appropriately for high-ability learners yet can remain accessible to other learners.
3. The curriculum framework provides a way for readers to get a snapshot view of the key emphases of the curriculum in direct relation to each other. The model also provides a way to traverse the elements individually through the continuum of grade levels.

Moreover, the framework may be used to implement the William and Mary units and to aid in new curriculum development based on science reform recommendations.

DOI: 10.4324/9781003238386-1

# Table 1
# Project Clarion Curriculum
# Framework for Science Units

| Goal | Student Outcomes<br>The student will be able to: |
|---|---|
| 1. Develop concepts related to understanding the world of science. | · Provide examples and salient features of various concepts.<br>· Classify various concepts.<br>· Identify counterexamples of various concepts.<br>· Create definitions or generalizations about various concepts. |
| 2. Develop an understanding of the macroconcept of change as applied to science content goals. | · Understand that change is everywhere.<br>· Demonstrate the impact of time on change.<br>· Articulate the nature of natural versus manmade change.<br>· Evaluate the nature of change (predictable or random) in selected phenomena. |
| 3. Develop knowledge of selected content topics in the life sciences. | · Determine that all plants and animals undergo changes in their life cycle.<br>· Investigate that as animals and plants grow they get larger according to a pattern.<br>· Conclude that animals are similar to their parents.<br>· State the basic needs of plants and animals.<br>· Articulate that animals need a suitable place to live.<br>· Articulate that plants need a place to grow.<br>· Conclude that thriving plants are plants that are doing very well in their environment.<br>· Understand that different animals have different body coverings.<br>· Understand that different animals have different appendages.<br>· Identify the different ways animals move.<br>· Classify animals. |
| 4. Develop interrelated science process skills. | · Make observations.<br>· Ask questions.<br>· Learn more.<br>· Design and conduct the experiment.<br>· Create meaning.<br>· Tell others what was found. |
| 5. Develop critical thinking skills. | · Describe problematic situations or issues.<br>· Define relevant concepts.<br>· Identify different points of view in situations or issues.<br>· Describe evidence or data supporting a scientific question.<br>· Draw conclusions based on data (making inferences).<br>· Predict consequences. |
| 6. Develop creative thinking skills. | · Develop fluency when naming objects and ideas, based on a stimulus.<br>· Develop flexible thinking.<br>· Elaborate on ideas presented in oral or written form.<br>· Create novel products. |
| 7. Develop curiosity and interest in the world of science. | · Express reactions about discrepant events.<br>· Ask meaningful questions about science topics.<br>· Articulate ideas of interest about science.<br>· Demonstrate persistence in completing science tasks. |

## Standards Alignment

Each lesson was aligned to the appropriate National Science Education Standards (NSES), Content Standards: K–4 (Center for Science, Mathematics, and Engineering Education [CSMEE], 1996). Table 2 presents detailed information on the alignment between the NSES Content Standards and fundamental concepts within the unit lessons.

# Table 2
## *Survive and Thrive* Alignment to National Science Education Standards

| Standard | Fundamental Concepts | Unit Lesson |
|---|---|---|
| Content Standard A: Abilities necessary to do scientific inquiry | · Ask a question about objects, organisms, and events in the environment.<br>· Plan and conduct a simple investigation.<br>· Employ simple equipment and tools to gather data and extend the senses.<br>· Use data to construct a reasonable explanation.<br>· Communicate investigations and explanations. | 1, 2, 3, 4, 5, 6, 7, 8, 9, 10, 11, 12 |
| Content Standard A: Understanding about scientific inquiry | · Scientific investigations involve asking and answering a question and comparing the answer with what scientists already know about the world.<br>· Scientists use different kinds of investigations depending on the questions they are trying to answer. Types of investigations include: describing objects, events, and organisms; classifying them; and doing a fair test (experimenting).<br>· Simple instruments, such as magnifiers, thermometers, and rulers, provide more information than scientists obtain using only their senses.<br>· Scientists develop explanations using observations (evidence) and what they already know about the world (scientific knowledge). Good explanations are based on evidence from investigations.<br>· Scientists make the results of their investigations public; they describe the investigation in ways that enable others to repeat the investigation.<br>· Scientists review and ask questions about the results of other scientists' work. | 1, 2, 3, 4, 5, 6, 7, 8, 9, 10, 11, 12 |
| Content Standard C: The characteristics of organisms | · Organisms have basic needs. For example, animals need air, water, and food; plants require air, water, nutrients, and light. Organisms can survive only in environments in which their needs can be met. The world has many different environments, and distinct environments support the life of different types of organisms.<br>· Each plant or animal has different structures that serve different functions in growth, survival, and reproduction. For example, humans have distinct body structures for walking, holding, seeing, and talking.<br>· The behavior of individual organisms is influenced by internal cues (such as hunger) and by external cues (such as change in the environment). Humans and other organisms have senses that help them detect internal and external cues. | 6, 7, 8, 9, 10, 12 |
| Content Standard C: Life cycles of organisms | · Plants and animals have life cycles that include being born, developing into adults, reproducing, and eventually dying. The details of this life cycle are different for different organisms.<br>· Plants and animals closely resemble their parents.<br>· Many characteristics of an organism are inherited from the parents of the organism, but other characteristics result from an individual's interactions with the environment. Inherited characteristics include the color of flowers and number of limbs of an animal. Other features, such as the ability to ride a bicycle, are learned through interactions with the environment and cannot be passed on to the next generation. | 5, 6, 12 |
| Content Standard C: Organisms and environments | · All animals depend on plants. Some animals eat plants for food. Other animals eat animals that eat the plants.<br>· An organism's patterns of behavior are related to the nature of that organism's environment, including the kinds and numbers of other organisms present, the availability of food and resources, and the physical characteristics of the environment. When the environment changes, some plants and animals survive and reproduce and others die or move to a new location.<br>· All organisms cause changes in the environment where they live. Some of these changes are detrimental to the organism or other organisms, whereas others are beneficial.<br>· Humans depend on their natural and constructed environments. Humans change environments in ways that can be either beneficial or detrimental for themselves and other organisms. | 7, 8, 11, 12 |

## Table 2, continued

| Standard | Fundamental Concepts | Unit Lesson |
|---|---|---|
| Content Standard F: Changes in environments | · Environments are the space, conditions, and factors that affect an individual's and a population's ability to survive and their quality of life.<br>· Changes in environments can be natural or influenced by humans. Some changes are good, some are bad, and some are neither good nor bad. Pollution is a change in the environment that can influence the health, survival, or activities of organisms, including humans.<br>· Some environmental changes occur slowly, and others occur rapidly. Students should understand the different consequences of changing environments in small increments over long periods as compared to changing environments in large increments over short periods. | 2, 11, 12 |

## Macroconcept

The macroconcept for this unit is *change*. A concept paper on change is included in Appendix A. The natural world changes continually; however, some changes may be too slow to observe. Students begin to understand the concept of change in science by learning about natural changes that occur over time, as well as manmade changes that impact conditions. The second lesson in this unit introduces the concept of change. Students are asked to brainstorm examples of change, categorize their examples, identify "nonexamples" of the concept, and make generalizations about the concept (Taba, 1962). The generalizations about change incorporated into this unit of study include:

- Change is everywhere.
- Change is related to time.
- Change can be natural or manmade.
- Change may be random or predictable.

The concept of change is integrated throughout the unit lessons and deepens students' understanding of animals and plants. Students examine the relationship of important ideas, abstractions, and issues through application of the concept generalizations. This higher-level thinking enhances the students' ability to "think like a scientist." More information about concept development is provided in Appendix B: Teaching Models.

## Key Science Concepts

By the end of this unit, students will understand that:
- All plants and animals undergo different changes in their life cycle.
- As animals and plants grow, they get larger according to a pattern.
- Animals are similar to their parents.
- Plants and animals have basic needs for oxygen, food, and water.
- Animals need a suitable place to live.
- Plants need a place to grow.
- Thriving plants are plants that are doing very well in their environment.
- Animals have different body coverings, such as hair, fur, feathers, scales, and shells.
- Body coverings help animals in different ways.
- Animals have different appendages, such as arms, legs, wings, fins, and tails.
- Animals move in different ways, such as walking, crawling, flying, climbing, or swimming.
- Animals can be classified in different ways.

Practice in using concept maps supports students' learning as they begin to build upon known concepts (Novak & Gowin, 1984). Students begin to add new concepts to their initial understandings of a topic and to make new connections between concepts. The use of concept maps within the lessons also helps teachers to recognize students' conceptual frameworks so that instruction can be adapted as necessary. More information on strategies for using concept mapping, as well as a list of concept-mapping practice activities, is provided in Appendix B: Teaching Models.

Each Project Clarion unit contains a science concept map (see Figure 1) that displays the essential understandings and the connections students should be able to make as a result of their experiences within the unit. This overview may be useful as a classroom poster that teachers and students can refer to throughout the unit.

## Scientific Investigation and Reasoning

The Wheel of Scientific Investigation and Reasoning contains the specific processes involved in scientific inquiry that guide students' thinking and actions. To read more about these processes and suggestions for implementing the wheel into this unit's lessons, see Appendix B.

The lessons that utilize the Wheel of Scientific Investigation and Reasoning include:

- Lesson 1, which helps students gain a better understanding of what scientists do and introduces the Wheel of Scientific Investigation and Reasoning, including the six components of scientific investigation.
- Lessons 3 and 4, which continue the introduction to scientific investigation by requiring students to make observations, ask questions, learn more about a topic, design and conduct experiments, create meaning, and tell others what was found.
- Lessons 5 to 11, which provide opportunities for students to use one or more components of scientific investigation and reasoning.

Students apply the components of scientific investigation throughout the unit and use the wheel to analyze aspects of an investigation or to plan an investigation. Scientific investigation concepts within the lessons include:

- *Make Observations*: Scientists use their senses as well as instruments to note details, identify similarities and differences, and record changes in phenomena.
- *Ask Questions*: Scientists use information from their observations about familiar objects or events to develop important questions that spark further investigation.
- *Learn More*: Scientists carefully review what is known about a topic and determine what additional information must be sought.
- *Design and Conduct the Experiment*: Scientists design an experiment, which is a fair test of a hypothesis or prediction and is intended to answer a question for a scientific investigation.
- *Create Meaning*: Scientists carefully gather and record data from an experiment, then analyze the data.
- *Tell Others What Was Found*: Scientists communicate findings from an experiment, including a clear description of the question, the hypothesis or prediction, the experiment that was conducted, the data that were collected and how they were analyzed, and the conclusions and inferences that were made from the experiment.

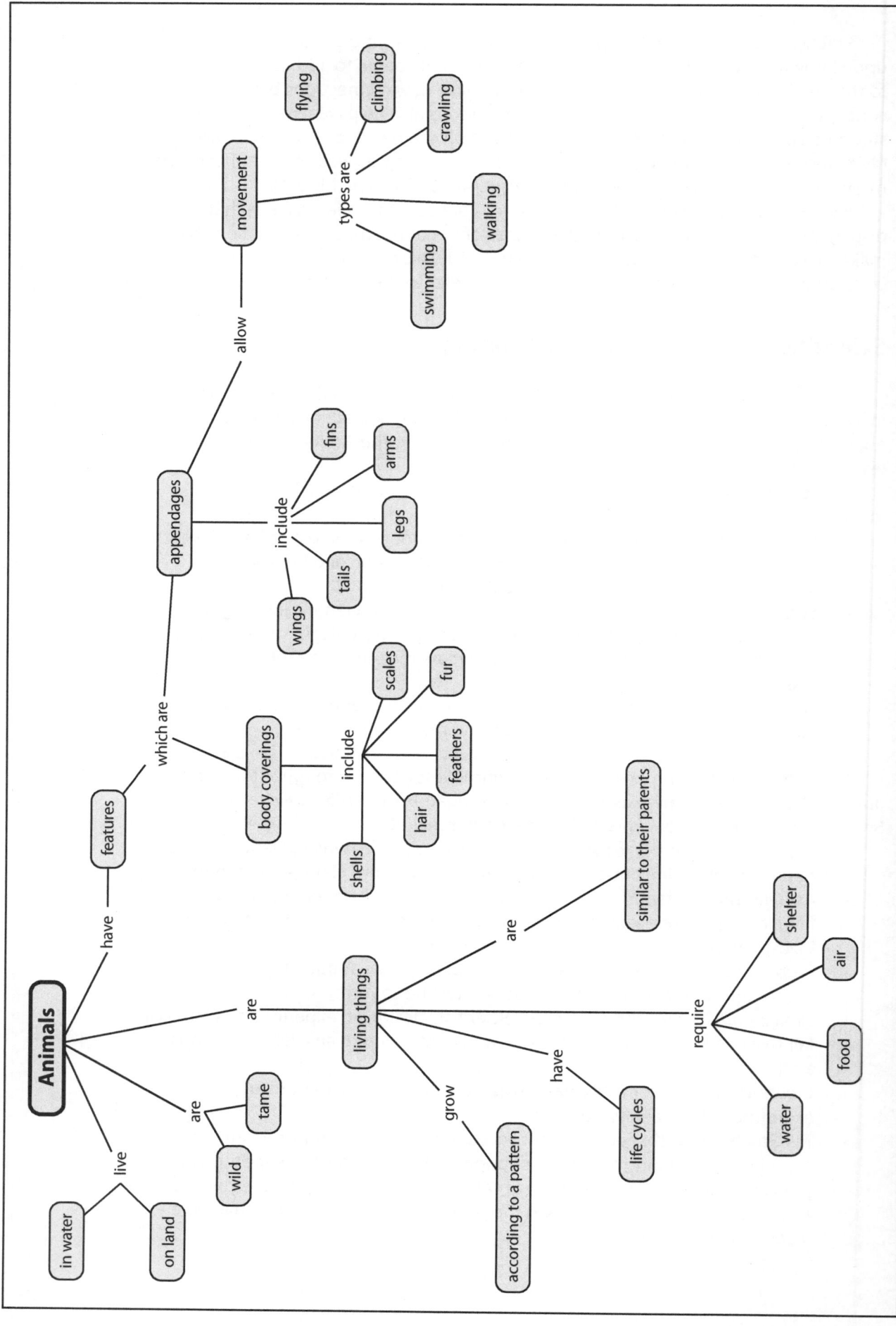

**Figure 1.** Unit concept map.

8

## Assessment

The unit includes performance-based assessments for students to complete at the beginning (preassessment) and end (postassessment) of the unit. There are three pre- and postassessments, which assess conceptual understanding, science content knowledge, and application of the scientific investigation process. The preassessment provides baseline data that teachers can use to adjust instructional plans for individual students or groups of students. Preteaching activities accompany selected preassessments.

The postassessment is administered at the completion of the unit and provides valuable information about students' mastery of the targeted objectives and the National Science Education Standards. A rubric is used to score each pre- and postassessment. The pre- and postassessments and dimensions of learning scored for each task include:

- *A macroconcept template, which requires students to draw or write about the macroconcept.* Conceptual understanding is scored on the pre- and postassessments based on the number of appropriate examples of the macroconcept, the elements of the macroconcept, types of the macroconcept listed, and generalizations about the macroconcept.
- *Concept maps, which assess students' content knowledge.* Students are given a prompt for creating a concept map about the unit topic. Understanding of key science concepts is scored on the pre- and postassessments based on the number of appropriate hierarchical levels, propositions, and examples listed.
- *An experimental design template, which requires students to plan an experiment with a given scientific research question.* Students are asked to design an experiment to investigate a question. Students are scored on the pre- and postassessments on their ability to write a prediction or hypothesis, list materials needed for the experiment, list the steps of the experiment in order, and develop a plan to organize data for collection and interpretation.

Teachers also should note that assessment "Look Fors" are designated in the first section of each lesson plan. The "Look Fors" provide a means for teachers to assess student learning in each lesson. The "Look Fors" are linked to the macroconcept generalizations, key science concepts, and scientific processes identified in each lesson. Teachers can develop checklists for the "Look Fors" or may make informal observations.

# Teacher's Guide to Content

The following definitions of key science concepts taught in the unit are described along with a unit glossary and a list of content resources.

## Survive and Thrive

Plants and animals have basic needs that must be met in order for them to survive. These basic needs include oxygen, food, water, and shelter or a place to grow. The animal population around the world is very diverse. Animals and plants either adapt to their environment and the changes that occur in it, or they die out. These adaptations may include changes to body coverings, appendages, or where the animal lives. Many animals have camouflage that helps them blend in with their environment. This can help an animal avoid a predator, or it can help a predator capture its prey.

Body coverings include hair, fur, feathers, scales, and shells. Body coverings provide protection from predators and the elements. They also keep the animal's body temperature even and can keep the animal dry. Animal appendages include arms, legs, wings, fins, and tails. These appendages enable animals to move in different ways, such as walking, crawling, flying, swimming, or climbing.

Animals can be classified or grouped according to certain characteristics. Animals can be tame or wild. Animals live on land or in water. For greater complexity, the Linnaean taxonomy can be used. It was first developed in the 1700s by a scientist named Carolus Linnaeus, although the form we use today is much more complex. In this taxonomic system, all life is classified into a tree-like hierarchy using various categories. The broadest category is the kingdom, followed by phylum, class, order, family, genus, and, finally, species.

When an animal or plant is thriving, it is doing well or flourishing. Kudzu, a vine that has grown out of control in the southeastern United States, is an example of a plant that is thriving. Likewise, deer, which have become overpopulated in many areas, causing damage to the environments in which they live, are an example of animals that are thriving. Endangered animals, such as the blue whale, are declining.

## Life Cycles

All plants and animals undergo changes in their life cycle. Plants and animals get larger according to a pattern. Animals inherit traits from their parents and are, therefore, similar to them.

All animal life cycles share certain common elements: an animal is born, grows and changes, ages, and eventually dies. However, there are important differences as well; a lion's life cycle is very different from a butterfly's. We can make generalizations about life cycles based on the classes of animals. We will look at the life cycles of insects, arachnids, amphibians, reptiles, and mammals to see how they are similar and different.

All insects hatch from eggs. Some insects look like a miniature, wingless version of the adult. This form is called a nymph. The nymph grows, and its wings develop outside of its body. After the wings have grown, the insect is an adult. Grasshoppers and cockroaches are examples of insects that grow from nymph to adult. Other insects, such as butterflies and houseflies, hatch from their eggs in an early form that

DOI: 10.4324/9781003238386-2

**10**

looks very different from their eventual adult form. This early stage is called a larva. Larvae are basically eating machines. When they have eaten enough, they find a safe place, molt their larval skin, and grow a new, thicker skin. This stage of the insect's life is known as a pupa. While in the pupal stage, the insect's body changes dramatically and takes on the adult form. This process is known as metamorphosis. When these changes are complete, the adult insect emerges from the pupa.

Arachnids, including spiders and scorpions, hatch from eggs into a nymph form like some insects. This nymph form has only six legs, although it otherwise looks like an adult. As the nymph grows, it molts its skin. When the nymph grows its fourth pair of legs and reaches sexual maturity, it is an adult, also called an imago.

Amphibians also hatch from eggs and undergo metamorphosis. The metamorphosis of an amphibian is particularly dramatic. Adult amphibians live on land, but they lay their eggs in water. The larval form (a frog tadpole) breathes through gills, has a tail, and eats plants. Slowly, the larva grows legs and lungs and loses its tail and gills. The mouth changes position on the body and grows larger. The amphibian moves out of the water to live on land and changes to its adult diet of insects, worms, and other small animals. This process can take as long as a year.

The life cycle of a reptile is not nearly so dramatic. Reptiles hatch from eggs looking like small versions of their adult parents. Some reptiles, such as snakes, shed their skins as they grow larger. Reptiles are considered adults when they reach sexual maturity.

Mammals are unique in that, rather than hatching from eggs, they are born alive. Like reptiles, baby mammals look much like their parents when they are born and are considered adults at sexual maturity.

# Unit Glossary

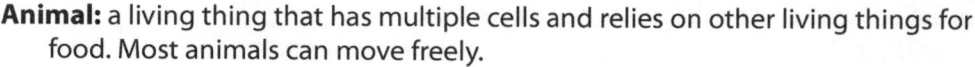

**Animal:** a living thing that has multiple cells and relies on other living things for food. Most animals can move freely.

**Appendage:** a body part such as an arm, leg, tail, or fin that is attached to an animal.

**Endangered:** a plant or animal that is under threat of becoming extinct.

**Habitat:** the place or natural environment in which an animal or plant lives.

**Life cycle:** the stages of growth and development that an organism goes through in its lifetime.

**Mealworm:** the larva stage of a grain beetle.

**Survive:** to continue to live.

**Tame:** brought or bred out of the wild state; domesticated.

**Thrive:** to grow heartily.

**Wild:** existing in a natural state; not tamed.

DOI: 10.4324/9781003238386-3

# Teaching Resources

## Required Resources (Used in Relevant Lessons)

Aliki. (1991). *My five senses.* New York, NY: HarperFestival.
Arlon, P. (2004). *First animal encyclopedia.* New York, NY: DK Publishing.
Lehn, B. (1999). *What is a scientist?* Brookfield, CT: Millbrook Press.
Schaffer, D. (1999). *Mealworms.* Mankato, MN: Capstone Press.
Scholastic. (2007). *Scholastic children's dictionary.* New York, NY: Scholastic.

## Additional Resources

Carle, E. (1981). *The very hungry caterpillar.* New York, NY: Philomel.
Carle, E. (2009). *The tiny seed.* New York, NY: Little Simon.
Conway, C. (Director). (1997). *Eyewitness: Pond and river* [Motion picture]. New York, NY: DK Video.
Fowler, A. (1998). *Hard-to-see animals.* New York, NY: Children's Press.
Gibbons, G. (1993). *From seed to plant.* New York, NY: Holiday House.
Hickman, P. (2000). *Animals in motion: How animals swim, jump, slither, and glide.* Tonawanda, NY: Kids Can Press.
Kalman, B. (2000). *How do animals adapt?* New York, NY: Crabtree.
Kaner, E. (1999). *Animal defenses: How animals protect themselves.* Tonawanda, NY: Kids Can Press.
*Kingfisher first encyclopedia of animals.* (2005). Boston, MA: Kingfisher.
LongNeedle Entertainment. (2008). *Animal atlas: Best of season 3, the complete set* [Motion picture]. Burbank, CA: LongNeedle Entertainment.
McGhee, K., and McKay, G. (2006). *National Geographic encyclopedia of animals.* Des Moines, IA: National Geographic Children's Books.
Stouffer, M. (Producer). (2008). *Wild America* [Television series]. Renton, WA: Topics Entertainment.
Wilk, A. C. (Producer). (1994). *National geographic: Really wild animals—Swinging safari* [Television series]. Washington, DC: National Geographic Video.
Wilk, A. C. (Producer). (1994). *National geographic: Really wild animals—Totally tropical rainforest* [Television series]. Washington, DC: National Geographic Video.
Wilkes, A. (2007). *Animal homes.* Boston, MA: Kingfisher.

## Useful Websites

Annenberg Media. (2010). *Journey north: A global study of wildlife migration and seasonal change.* Retrieved from http://www.learner.org/jnorth
Big Cat Rescue. *Welcome to Big Cat Rescue!* Retrieved from http://www.bigcatrescue.org
Kidport Reference library. (n.d.). *Animal homes.* Retrieved from http://www.kidport.com/RefLib/Science/AnimalHomes/AnimalHomes.htm
Monterey Bay Aquarium. (n.d.). *Podcasts, videos, and web cams.* Retrieved from http://www.mbayaq.org/efc/cam_menu.asp
Museum Victoria Australia. (2002). *The spider's parlour: Tarantula web-cam.* Retrieved from http://www.museum.vic.gov.au/spidersparlour/tarant.htm

 DOI: 10.4324/9781003238386-4

National Geographic Society. (n.d.) *Crittercam chronicles*. Retrieved from http://www.nationalgeographic.com/crittercam/index.html

Public Broadcasting Service. (2001). *American field guide*. Retrieved from http://www.pbs.org/americanfieldguide/topics/animals/index.html

San Diego Zoo. (2010). *Panda canyon*. Retrieved from http://www.sandiegozoo.org/zoo/animal_zones/panda_canyon/panda_exhibit

San Diego Zoo. (2010). *Polar bear plunge*. Retrieved from http://www.sandiegozoo.org/polarbearplunge

Smithsonian National Zoological Park. (n.d.). *AnimalCams*. Retrieved from http://nationalzoo.si.edu/Animals/WebCams

U.S. Fish and Wildlife Service. (2009). *Endangered species program: Kid's corner*. Retrieved from http://www.fws.gov/endangered/kids/index.html

U.S. Fish and Wildlife Service. (2010). *Species*. Retrieved from http://www.fws.gov/species

Wild Birds Unlimited. (2010). *Operation migration cranecam*. Retrieved from http://www.wbu.com/feedercam_home.html

World Wildlife Fund. (2010). *Gift center*. Retrieved from http://www.worldwildlife.org/forms/adoptionCenter_1.cfm?donorsrc=leftnav&winter05=leftnav

# Part II: Lesson Plans

## Lesson Plans

Overview of Lessons

Lesson 1: What Is a Scientist?

Lesson 2: What Is Change?

Lesson 3: What Scientists Do—Observe, Question, Learn More

Lesson 4: What Scientists Do—Experiment, Create Meaning, Tell Others

Lesson 5: What Is a Life Cycle?

Lesson 6: What Is the Life Cycle of a Mealworm?

Lesson 7: What Are the Requirements for Life?

Lesson 8: What Do Plants Need to Thrive?

Lesson 9: How Do Animals Look Different?

Lesson 10: What Is an Appendage?

Lesson 11: How Can We Classify Animals?

Lesson 12: What Have We Learned About Animals and Plants?

Postassessment

DOI: 10.4324/9781003238386-5

# Overview of Lessons

An overview of the lessons is provided in Table 3. The overview shows the primary emphasis of each lesson in the unit according to the macroconcept, key science concepts, or the scientific investigation process. Lessons also may have a secondary emphasis, which is listed in the planning section of each lesson, labeled "Planning the Lesson."

## Table 3
## Overview of Lessons

| Concept of Change | Scientific Process | Key Science Concepts |
|---|---|---|
| Preassessment | | |
| | Lesson 1: What Is a Scientist? | |
| Lesson 2: What Is Change? | | |
| | Lesson 3: What Scientists Do—Observe, Question, Learn More | |
| | Lesson 4: What Scientists Do—Experiment, Create Meaning, Tell Others | |
| | | Lesson 5: What Is a Life Cycle? |
| | Lesson 6: What Is the Life Cycle of a Mealworm? | |
| | | Lesson 7: What Are the Requirements for Life? |
| | Lesson 8: What Do Plants Need to Thrive? | |
| | | Lesson 9: How Do Animals Look Different? |
| | | Lesson 10: What Is an Appendage? |
| | Lesson 11: How Can We Classify Animals? | |
| Lesson 12: What Have We Learned About Animals and Plants? | | |
| Postassessment | | |

## Lesson Plan Blueprint

The lesson plan blueprint for each lesson shows:
- the instructional purpose,
- generalizations about the macroconcept of change,
- key science concepts,
- scientific investigation skills and processes, and
- assessment "Look Fors."

# Table 4
# Lesson Plan Blueprint

| Lesson # | Title | Instructional Purpose | Change Generalizations | Key Science Concepts | Scientific Investigation Skills and Processes | Assessment "Look Fors" Students should be able to: |
|---|---|---|---|---|---|---|
| | Preassessment | | | | | |
| 1 | What Is a Scientist? | • To learn the characteristics of scientists and the investigation skills that scientists use. | | | • Make observations<br>• Ask questions<br>• Learn more<br>• Design and conduct the experiment<br>• Create meaning<br>• Tell others what was found | • Identify the scientific investigation processes used by scientists. |
| 2 | What Is Change? | • To understand the concept of change.<br>• To learn four generalizations about change that will deepen understanding of animals. | • Change is everywhere.<br>• Change is related to time.<br>• Change can be natural or manmade.<br>• Change may be random or predictable. | | • Make observations<br>• Learn more | • Give examples of things that change.<br>• Categorize examples of change, explaining their reasoning.<br>• Show understanding of the concept of "generalization."<br>• Give an example of a change generalization. |
| 3 | What Scientists Do—Observe, Question, Learn More | • To apply three of six investigation processes (make observations, ask questions, and learn more) described in the Wheel of Scientific Investigation and Reasoning.<br>• To initiate an investigation of flowers. | • Change is everywhere. | | • Make observations<br>• Ask questions<br>• Learn more<br>• Design and conduct the experiment<br>• Create meaning<br>• Tell others what was found | • Apply the scientific investigation process. |

*Table 3, continued*

| Lesson # | Title | Instructional Purpose | Change Generalizations | Key Science Concepts | Scientific Investigation Skills and Processes | Assessment "Look Fors" Students should be able to: |
|---|---|---|---|---|---|---|
| 4 | What Scientists Do—Experiment, Create Meaning, Tell Others | • To design and conduct an experiment using flowers. <br>• To create meaning from the experiment. <br>• To learn how to tell others what was found. | • Change is everywhere. <br>• Change is related to time. <br>• Change can be natural or manmade. <br>• Change may be random or predictable. | | • Make observations <br>• Ask questions <br>• Learn more <br>• Design and conduct the experiment <br>• Create meaning <br>• Tell others what was found | • Design and conduct an experiment. <br>• Interpret data from a data table. <br>• Describe how the experiment was conducted and what results were found. |
| 5 | What Is a Life Cycle | • To demonstrate an understanding that all plants and animals undergo different changes as they grow and develop. | • Change is everywhere. <br>• Change is related to time. <br>• Change may be random or predictable. | • All plants and animals undergo different changes in their life cycles. <br>• As animals and plants grow; they get larger according to a pattern. <br>• Animals are similar to their parents. | • Make observations. <br>• Ask questions. <br>• Learn more. | • Express the idea that living things grow and change. <br>• Explain the life cycle of living things. |
| 6 | What Is the Life Cycle of a Mealworm? | • To investigate the life cycle of a mealworm. | • Change is related to time. <br>• Change can be natural or manmade. <br>• Change may be random or predictable. | • Plants and animals undergo different changes in their life cycles. <br>• Animals are similar to their parents. | • Make observations <br>• Ask questions <br>• Learn more <br>• Design and conduct the experiment <br>• Create meaning <br>• Tell others what was found | • Give reasons to support their hypotheses. <br>• Describe the life cycle of a mealworm. <br>• Record observations on a data table. <br>• Report findings on the Experimental Report Form. |
| 7 | What Are the Requirements for Life? | • To identify the basic needs of all animals: oxygen, food, water, and shelter or a place to grow. | • Change can be natural or manmade. | • Plants and animals have basic needs for oxygen, food, and water. <br>• Animals need a suitable place to live. | • Make observations <br>• Ask questions <br>• Learn more | • State the basic needs of animals. |
| 8 | What Do Plants Need to Thrive? | • To understand the needs of plants. <br>• To design an experiment using the Wheel of Scientific Investigation and Reasoning. <br>• To understand the difference between surviving and thriving in plants. | • Change is related to time. <br>• Change can be natural or manmade. | • Plants and animals have basic needs for oxygen, food, and water. <br>• Plants need a place to grow. <br>• Thriving plants are plants that are doing very well in their environment. | • Ask questions <br>• Design and conduct the experiment <br>• Create meaning | • Identify the needs of plants for survival. <br>• Distinguish between surviving and thriving. |

| Lesson # | Title | Instructional Purpose | Change Generalizations | Key Science Concepts | Scientific Investigation Skills and Processes | Assessment "Look Fors" Students should be able to: |
|---|---|---|---|---|---|---|
| 9 | How Do Animals Look Different? | • To identify the different animal body coverings, such as hair, fur, feathers, scales, and shells, and their purpose. | • Change is related to time. | • Animals have different body coverings, such as hair, fur, feathers, scales, and shells.<br>• Body coverings help animals in different ways. | • Make observations<br>• Learn more<br>• Tell others what was found | • Identify body coverings.<br>• Identify which body coverings go with which animals.<br>• Identify a purpose for body coverings. |
| 10 | What Is an Appendage? | • To understand how animal movement occurs. | • Change is everywhere. | • Animals have different appendages, such as arms, legs, wings, fins, and tails.<br>• Animals move in different ways, such as walking, crawling, flying, climbing, or swimming. | • Make observations | • Match appendages to the correct animals.<br>• Identify which appendages match which movements. |
| 11 | How Can We Classify Animals? | • To classify animals in different ways. | • Change is everywhere. | • Animals can be classified in different ways. | • Make observations<br>• Create meaning | • Classify animals in different ways. |
| 12 | What Have We Learned About Animals and Plants? | • To apply what has been learned about animals and plants surviving and thriving. | | • All plants and animals undergo different changes in their life cycle.<br>• As animals and plants grow, they get larger according to a pattern.<br>• Animals are similar to their parents.<br>• Plants and animals have basic needs for oxygen, food, and water.<br>• Animals need a suitable place to grow.<br>• Plants need a place to grow.<br>• Thriving plants are plants that are doing very well in their environment.<br>• Animals have different body coverings, such as hair, fur, feathers, scales, and shells.<br>• Body coverings help animals in different ways.<br>• Animals have different appendages, such as arms, legs, wings, fins, and tails.<br>• Animals move in different ways, such as walking, crawling, flying, climbing, or swimming.<br>• Animals can be classified in different ways. | • Make observations<br>• Ask questions<br>• Learn more<br>• Design and conduct the experiment<br>• Create meaning<br>• Tell others what was found | • Identify the basic needs of animals and plants.<br>• Match body coverings to groups of animals.<br>• Match appendages with movement.<br>• Classify animals. |
| | Postassessment | | | | | |

# Preteaching Lesson: Science Safety

## Planning the Lesson

### Instructional Purpose

- To instill in students the importance of safety in the classroom.
- To outline science safety rules to be implemented throughout the unit.

### Instructional Time
- 45 minutes

### Materials/Resources/Equipment
- Sample materials:
  - Plant
  - Plastic bag of nonhazardous powdery substance (e.g., sugar)
  - Closed jar of nonhazardous liquid (e.g., water)

- Plastic disposable gloves
- Safety goggles
- Chart paper
- Markers
- Science Safety Guidelines (Handout 0A)
- Science Safety Rules printed on chart paper (Handout 0B)

## Implementing the Lesson

1. Display sample materials on a long table in front of students. Inform students that they soon will begin a science unit in which they will observe and study many different kinds of materials, such as these. Explain that it is important for students to practice safety during the investigations. Relate the necessity of science safety rules to those of the classroom and in physical education.
2. Display and define each item. Tell students that, as a class, they will create a list of rules they should follow when handling these materials. Have students think about how they can keep their bodies safe. Record these examples on chart paper.
3. Next, unveil the Science Safety Rules (Handout 0A) on chart paper. Have students compare the two lists. How do students' examples relate to these rules? If necessary, add additional rules to the list.
4. Explain why some materials (such as knives) or elements (such as fire) are never appropriate for children to handle in school. Briefly discuss the potential hazards associated with these materials.
5. Finally, conduct a brief demonstration to illustrate how to practice safety guidelines. Take the plastic bag containing a nonhazardous powdery substance and the jar of nonhazardous liquid. Explain that you are going to investigate how the two materials interact. Ask students how you can be safe while doing this investigation. Reinforce that substances can be harmful to the

eyes or skin and that they should **never** be ingested. Explain that the same is true of plants, which can be toxic to humans. Emphasize that students should follow similar guidelines when studying plants in other science units.

6. Following students' examples of safety measures, demonstrate how to use safety goggles to protect the eyes, plastic gloves to protect the hands, and other relevant protective measures, such as pulling long hair back and wearing appropriate clothing. Conduct the demonstration by carefully pouring the powdery substance into the jar of liquid. Emphasize that you should never touch your face or mouth (and especially should not eat or drink) during science experiments.

7. Tell students that materials will be disposed of properly by the teacher after the investigation is completed. Students should not touch any potentially harmful substances.

8. Demonstrate the final rule, "Wash your hands," by properly removing the gloves (without the outside of the gloves ever touching the body) and the goggles. If there is a sink in the classroom, demonstrate how to properly wash one's hands. If no sink is present, inform students that after each investigation the class will go to the bathroom to wash their hands.

9. Conclude the lesson by emphasizing that science investigations are interesting and fun, but they also can be dangerous if not conducted properly. By following the Science Safety Rules, the class will enjoy the benefits of learning about science.

# Handout 0A
## Science Safety Guidelines

1. Know and follow your school's policies and procedures regarding classroom safety.

2. Always provide direct adult supervision when students are engaging in scientific experimentation.

3. Ensure that all materials and equipment are safe for handling by primary students.

4. Exert extra caution when materials have the potential for harm when used improperly.

5. Use protective gear for eyes, skin, and breathing when conducting experiments, and require students to do the same.

6. Always conduct an experiment by yourself before completing it with the students.

7. Store materials for experiments out of the reach of students.

8. Never allow students to eat or drink during science experiments.

9. Follow general safety rules for sharp objects, heated items, breakables, or spilled liquids.

10. Teach students that it is unsafe to touch their face, mouth, eyes, or other body parts when they are working with plants, animals, microorganisms, or chemicals. Wash hands prior to touching anything. Caution students about putting anything in their mouth or breathing in the smell of substances.

11. Be aware of students' allergies to plants (including plant pollen) animals, foods, chemicals, or other substances to be used in the science classroom. Take all precautions necessary. Common food allergens include peanuts, tree nuts (cashews, almonds, walnuts, hazelnuts, macadamia nuts, pecans, pistachios, and pine nuts), shellfish, fish, milk, eggs, wheat, and soy.

12. Use caution with plants. Never allow students to pick or handle any unknown plants, leaves, flowers, seeds, or berries. Use gloves to touch unknown plants. Many common house, garden, and wooded area plants are toxic.

13. Avoid glass jars and containers. Use plastic, paper, or cloth containers.

14. Thermometers should be filled with alcohol, not mercury.

15. Clearly label any chemicals used and dispose of properly.

16. Teach students safety rules for science (see Handout 0B), including:
    a. **Always** do scientific experiments with an adult present.
    b. **Never** mix things together (liquids, powders) without adult approval.
    c. **Use** your senses carefully. Protect your eyes, ears, nose, mouth, and skin.
    d. **Wash your hands** after using materials for an experiment.

# Science Safety Rules

**1** **Always** do scientific experiments with an adult present.

**2** **Never** mix things together (liquids, powders) without adult approval.

**3** **Use** your senses carefully. Protect your eyes, ears, nose, mouth, and skin.

**4** **Wash your hands** after using materials for an experiment.

# Preassessment

## Planning the Lesson

### Instructional Purpose
- To determine prior knowledge of unit content.
- To build understanding of the unit macroconcept, science content, and science processes.

### Instructional Time
- Macroconcept assessment: 20 minutes
- Science content assessment: 30 minutes, including preteaching activity
- Scientific process assessment: 20 minutes

### Materials/Resources/Equipment
- Copies of Preassessment for Change Concept, Word Bank for Animals Concept Map, and Incomplete Animals Concept Map for each student
- Preassessment for Key Science Concepts, Rubric 1 (Scoring Rubric for Change Concept), Preteaching for Key Science Concepts Preassessment, Sample Concept Map, Rubric 2 (Scoring Rubric for Content Assessment), and Rubric 3 (Scoring Rubric for Scientific Process) for your use
- Copies of Does Sand Dissolve in Water?, What Materials Will You Need?, How Would You Conduct Your Experiment?, What Does This Table Show?, and What Will Dissolve? handouts for the Preassessment for the Science Process for each student
- Pencils
- Large chart paper
- Drawing paper for each student

## Implementing the Lesson

1. Each assessment should be administered on a different day.
2. Explain to students that the class is beginning a new unit of study. Tell them that they will be completing a preassessment to determine what they already know about the topic. Assure them that the assessment is not for a grade and encourage them to do their best.
3. Collect all of the preassessments. Briefly review each assessment and discuss some of the responses in general, indicating that this unit will provide them with more knowledge and skills than they now have.

## Scoring
- Score the preassessments using the rubrics provided. Keep the scores and assessments for diagnostic purposes to organize groups for various activities during the unit and to compare pre- and postassessment results.

DOI: 10.4324/9781003238386-7

Name:_____ Date:_____

# Preassessment for Change Concept

1. What is change? In each box, draw a picture or write a word for something that changes.

| | |
|---|---|
| | |
| | |
| | |
| | |
| | |

2. Draw a picture of something in your life that changes and show how it changes. Include as many details as you can.

3. Draw five ways a tree could change or be changed.

Name: _____     Date: _____

# Rubric 1

# Scoring Rubric for Change Concept

**Directions for Use:** Score students on their responses to each of the questions.

| | | 5 | 4 | 3 | 2 | 1 | 0 |
|---|---|---|---|---|---|---|---|
| 1 | **Examples of the Concept** | At least 9–10 appropriate examples are given. | At least 7–8 appropriate examples are given. | At least 5–6 appropriate examples are given. | At least 3–4 appropriate examples are given. | At least 1–2 appropriate examples are given. | No examples are given. |
| 2 | **Drawing of Before-After** | The drawing contains 5 picture elements that depict a before-after situation. | The drawing contains 4 picture elements that depict a before-after situation. | The drawing contains 3 picture elements that depict a before-after situation. | The drawing contains 2 picture elements that depict a before-after situation. | The drawing contains only 1 picture element that depicts a before-after situation. | The drawing contains no elements that depict a before-after situation. |
| 3 | **Types of Change** | Identifies 5 different types of change. | Identifies 4 different types of change. | Identifies 3 different types of change. | Identifies 2 different types of change. | Identifies 1 different types of change. | Identifies no type of change. |

**Total possible points: 15**

# Preteaching for Key Science Concepts Preassessment

**Directions for the Teacher:** Say the following bolded directions to students. Directions for you are not bolded.

**Sometimes we know a lot about something even before our teachers teach it in school. Sometimes we don't know very much at all, but we like to learn new things.**

**For example, what would you think about if someone asked you to tell all you know about how *farms* work? What are some of the words you would use?**

(List these on a chart.)

**What are some of the things that happen on a farm?**

(List these on a chart.)

**I am going to show you a way I might tell all I know about how farms work.**

(Begin a concept map on a large sheet of paper, using pictures and words, making simple links, and emphasizing these links.)

**Practice making your own concept map about a farm on your drawing paper. This practice activity can be done with a partner.** (Note to teacher: You may also choose to assign another familiar topic for the students to practice concept mapping.)

(Share some of the resulting concept maps, encouraging students to articulate their links.)

# Preassessment for
# Key Science Concepts

**Directions to the Teacher:** Read the following paragraph to the students.

Today I would like you to think about all of the things you know about animals. Think about the connections you can make. You will be completing a concept map, just like the ones you did when we discussed the farm. Look at the word bank and the concept map. You will use some of the word bank words to fill in the parts of the concept map. Some words are just extras that you won't need. Remember, a concept map is used to tell what we know and to make connections. Today's question is: "Tell me everything you know about animals."

## For Kindergarten Students

Direct students to use the word bank to complete the assessment. Students also may use other responses that they come up with on their own. Tell students to draw a picture or write the word or letter for their responses in the appropriate blanks. Each correct response earns one point. Students may enter the word *or* just the letter corresponding to the word *or* come up with their own word.

## For First-Grade Students

Direct students to complete the assessment with appropriate words, pictures, or their own choices of words. Each correct response earns one point.

Name: _____ Date: _____

# Preassessment for Key Science Concepts
# Incomplete Animals Concept Map

Some different animals are:

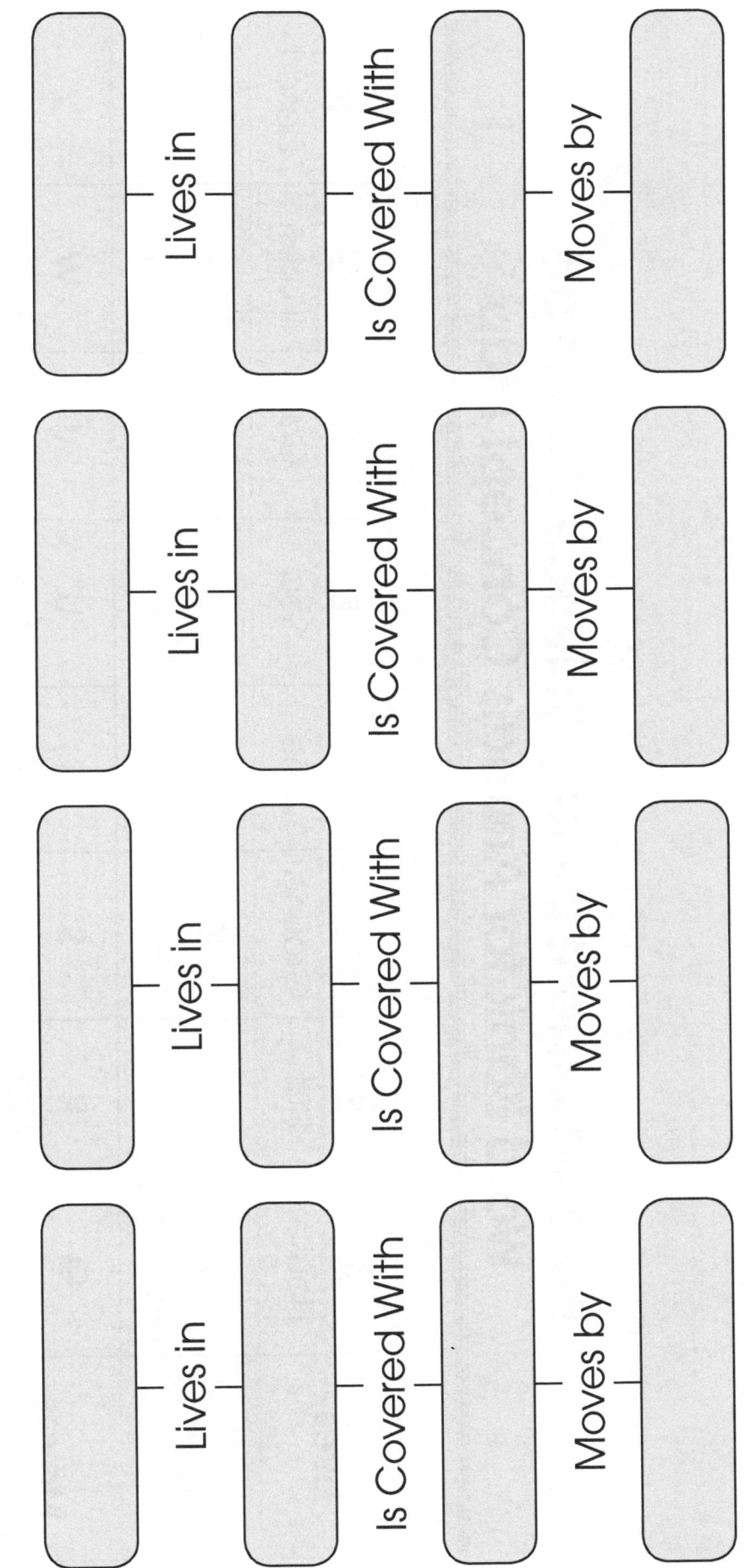

# Preassessment for Key Science Concepts
## Word Bank for Animals Concept Map

| Lives | A<br>House | B<br>Hut | C<br>Ground | D<br>Tree | E<br>Sea | F<br>Car | G<br>Forest | H<br>Desert |
|---|---|---|---|---|---|---|---|---|
| Covering | I<br>Feathers | J<br>Fly | K<br>Hair | L<br>Leaves | M<br>Scales | N<br>Shells | O<br>Fur | P<br>Water |
| Moves | Q<br>Swim | R<br>Walk | S<br>Scales | T<br>Fly | U<br>Tail | V<br>Crawl | W<br>Climb | X<br>Fin |

# Rubric 2
# Scoring Guide for Preassessment
# for Key Science Concepts

**Directions for Use:** Score students on their completed maps.

## Criteria

Concept Map (up to 16 points)

Score 1 point for each correct response in the concept map. Note that the student may have chosen a word that is not in the word bank. Score 1 point as long as the word(s) complete the link accurately. Students also may receive a point for each picture that accurately completes a link.

**Total possible points: 16**

# Preassessment for
# Scientific Process

1. Assess students in groups of 4 to 6.
2. Tell students they are going to think like scientists. Say to students, "I have a scientific question for you: Does sand dissolve in water? You are going to think about whether or not sand dissolves in water." We will work together to look at some pictures and select an answer to some questions about an experiment to find out if sand dissolves in water.
3. Pass out the packet of assessment record sheets on pp. 35–39. Ask students to look at the first sheet (Does Sand Dissolve in Water?). Ask them to write their name on the paper. Direct them to think about the two pictures and make a prediction about whether or not sand dissolves in water. Tell students to put an X in the box under the picture that shows their prediction—sand does not dissolve in water or sand does dissolve in water.
   Picture choices are:
   a.  Clear container with water and sand on the bottom
   b.  Clear container with water and no sand on the bottom

4. Ask students to think about what materials they will need for their experiment. Look at the What Materials Will You Need? handout (the one that shows some materials that could be used). Ask students to put an X under each picture that shows a material that will be used in the experiment.
   Picture choices are:
   a.  Clear container
   b.  Spoon
   c.  Sand
   d.  Salt
   e.  Water
   f.  Milk

5. Present each student with a set of four cards showing pictures of the steps in the experiment (see the How Would You Conduct the Experiment? handout). Tell the students to select the pictures that show the steps they would take for the experiment. Picture choices are (1) gathering the materials, (2) pouring in water, (3) pouring in sand, (4) stirring the mixture. Instruct students to put the steps they selected in the correct order—which comes first, second, etc. Check to see each student's response and record.
6. Ask students to look at the table on the What Does This Table Show? handout and decide whether it shows that sand dissolves in water or salt does not dissolve in water. Students should put an X in the correct box.
7. Ask students to look at the handout, What Will Dissolve?, with pictures of various materials. The materials are: leaf, twig, salt, JELL-O, crayon, sugar, rock, oatmeal. Direct students to think about things that probably dissolve in water and to place X in each box under a picture that shows something that will dissolve. Which of these materials will dissolve?

# Preassessment:
# Does Sand Dissolve in Water?

Does sand dissolve in water? Put an X in the box that matches your prediction.

# What Materials Will You Need?

What materials do you need to conduct your experiment? Put an X in the box of each material you would use.

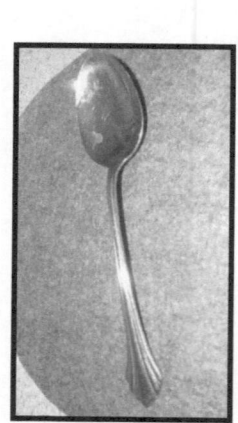

# How Would You Conduct
# Your Experiment?

Cut out the pictures below and place them in order of the steps of the water and sand experiment.

# What Does This Table Show?

Did the sand dissolve in water?

| | |
|---|---|
| Cassidy | No |
| Lowell | No |
| Sandy | No |
| Adrian | No |
| Leslie | No |
| Lincoln | No |
| Jonah | No |
| Chwee | No |
| Sun | No |

___ Yes, the sand dissolved in water.

___ No, the sand did not dissolve in water.

# What Will Dissolve?

Put an X in the box below each picture that shows something that will dissolve in water.

☐

☐

☐

☐

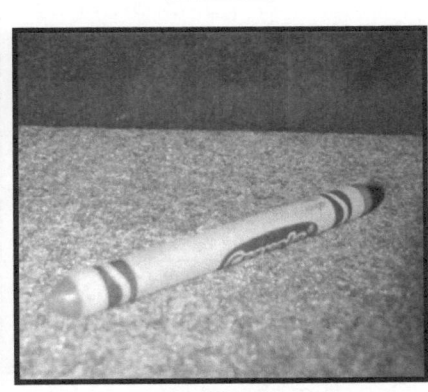
☐

☐

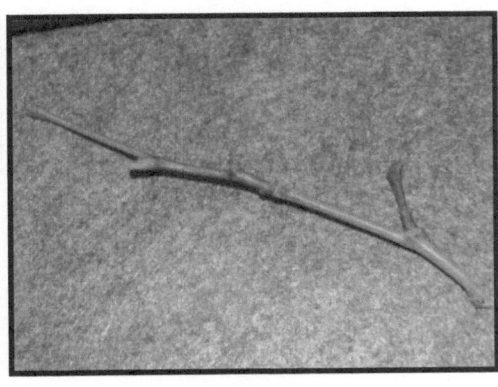
☐

☐

# Rubric 3
# Scoring Guide for Preassessment for Scientific Process

**Directions for Use:** Score students on the responses to each of the questions.

| Criteria | Scoring Guide | Points Available |
|---|---|---|
| Selects a prediction: Does sand dissolve in water? | Score the two glasses of water/sand as being either correct or incorrect. Give 1 point if the student selects the glass with sand at the bottom. | 1 point |
| Selects materials needed. | A total of up to 6 points is earned for checking the sand, empty glass, spoon, and water, and not checking salt or milk. | up to 6 points |
| Sequences steps. | One point is earned for each picture in its appropriate order. A total of up to 4 points may be given. The following order is correct: (1) the girl sitting with the ingredients in front of her, (2) the girl pouring water, (3) the girl pouring sand, and (4) the girl mixing. | up to 4 points |
| Selects the appropriate interpretation of the data table provided. | Two points awarded for the "No" response. | 2 points |
| Selects a prediction: What will dissolve? | One point should be given for each of the boxes being checked or not checked accurately. Checks should appear for salt, JELL-O, and sugar. No checks should appear for the remaining items. Students earn three points for each box they check correctly. | up to 9 points |

**Total possible points: _____/22**

# Lesson 1:
# What Is a Scientist?

## Planning the Lesson

### Instructional Purpose
- To learn the characteristics of scientists and the investigation skills that scientists use.

### Instructional Time
- 45 minutes

### Scientific Investigation Skills and Processes
- Make observations.
- Ask questions.
- Learn more.
- Design and conduct the experiment.
- Create meaning.
- Tell others what was found.

### Assessment "Look Fors"
- Students should be able to identify the scientific investigation processes used by scientists.

### Materials/Resources/Equipment
- Lab coat for teacher
- One lab coat (white adult T-shirt or dress shirt) for each student
- Beaker
- Microscope or magnifying glass
- Prepared charts for students, PowerPoint slides, or transparencies of Handouts 1A (Defining Scientists) and 1B (What Scientists Do: The Wheel of Scientific Investigation and Reasoning)
- Poster of the Wheel of Scientific Investigation and Reasoning
- Marker
- One piece of chart paper
- Student log books
- *What Is a Scientist?* by Barbara Lehn

> **Note to Teacher:** Scientific investigation is introduced to the students in this lesson and applied throughout the unit.

## Implementing the Lesson

1. Put on a lab coat and pick up a beaker and microscope or magnifying glass. Ask the students what kind of job you might have. Explain that you are a scientist. Ask the students if they know a scientist and allow them to discuss what they know about scientists or their experiences with scientists. Record student responses to the following questions (you may wish to refer to the Frayer Model of Vocabulary Development in Appendix B):
   - Do you know a scientist?
   - What do you think scientists do?

   DOI: 10.4324/9781003238386-8

2. Define a scientist as "a person who studies nature and the physical world by testing, experimenting, and measuring" (Scholastic, 2007) using Handout 1A.
3. Ask the students what they think scientists do and write down their responses on chart paper. Display the chart showing Handout 1B. Read the wheel to the students and talk about what each item means. Ask the students to compare the "What Scientists Do" processes with the list the class created.
4. Show students the book *What Is a Scientist?* by Barbara Lehn. Ask students to look for what scientists do while you are reading the book. Read the book, showing the pictures to the students and pointing out the clues that will help students understand what a scientist does. As you read each page, relate the activity to the scientific investigation processes included on the wheel. Ask students these questions:
   - What did the scientists do in the book?
   - What makes someone a scientist?
   - When is someone *not* a scientist?

5. Explain that the students will be working as scientists in the unit. Have students put on their "lab coats." Explain to the students that they are going to learn to think like a scientist and learn how to do what scientists do.
6. Tell students that scientists keep a scientific investigation log of what they are doing. They date the pages in their logs and then write down what they have learned or what they are thinking about what they learned. Tell students that they are going to keep a log and they are going to make the first page. Pass out student log books. Ask each student to date the first page and to draw a picture of him- or herself investigating something.
7. Have students share their completed pictures with the class.

## Concluding and Extending the Lesson

### Concluding Questions and/or Actions
- Would you like to be a scientist? Why or why not?
- All science is about how things stay the same and how they change. How do scientists study change?
- Set up a learning center where students can cut pictures out of magazines and newspapers of people who are scientists and paste the pictures on a class collage.
- Provide props (e.g., magnifying glasses) and lab coats for students to practice being scientists in the science investigation center.
- Provide books of individuals who are investigating something in the library center of the classroom.

### What to Do at Home
- Ask the students to ask an adult family member or some other adult they know to respond to the question, "What would you investigate/study/do if you were a scientist?"

# Defining Scientists

## A scientist is someone who . . .

studies nature and the physical world by testing, experimenting, and measuring.
(Scholastic, 2007)

## Scientists . . .

try to find answers to questions they have about our world. Often, they improve our world by finding answers to their questions.

# What Scientists Do: The Wheel of Scientific Investigation and Reasoning

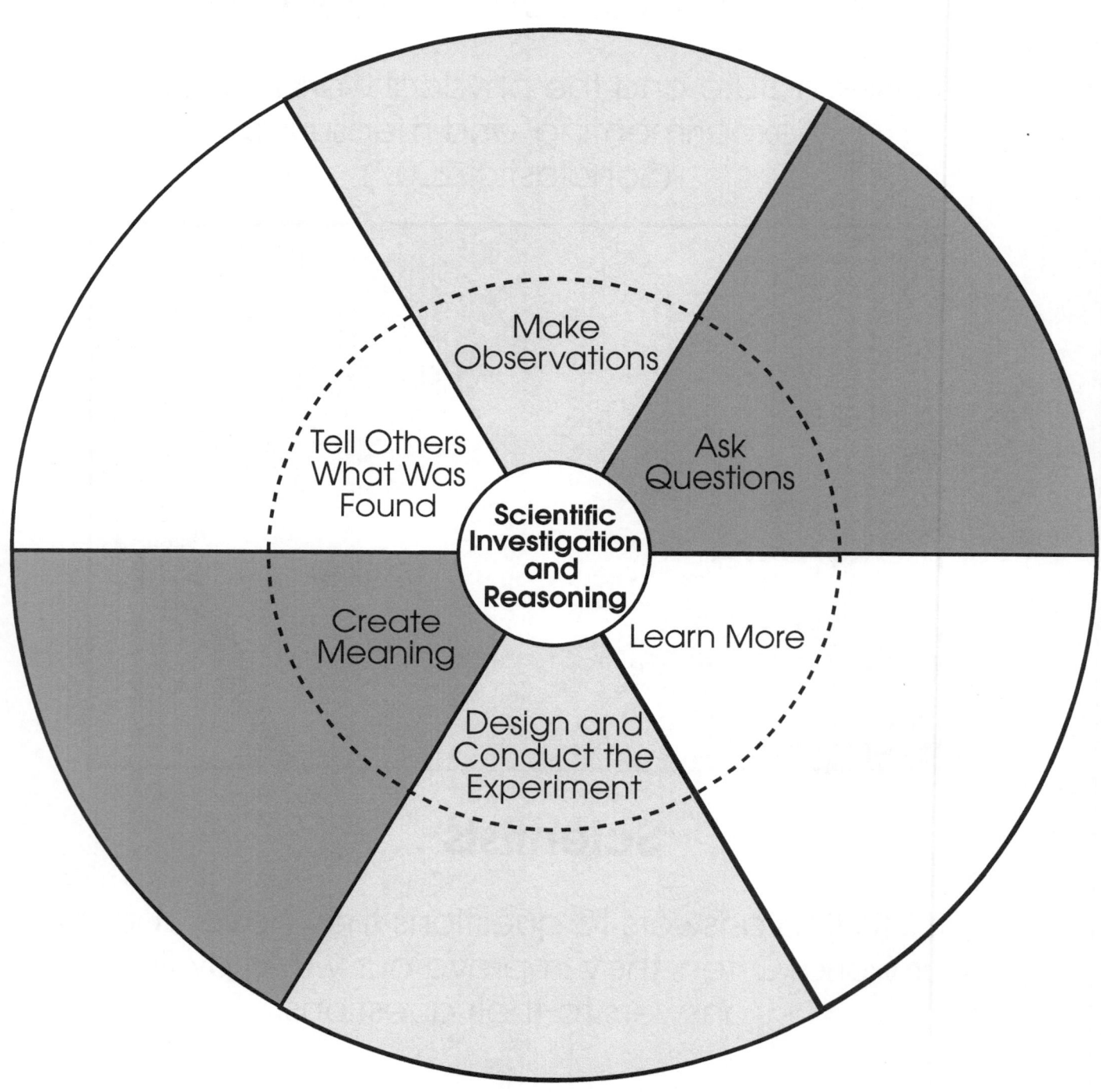

# Lesson 2:
# What Is Change?

## Planning the Lesson

### Instructional Purpose
- To understand the concept of change.
- To learn four generalizations about change that will deepen understanding of animals.

### Instructional Time
- 45 minutes

### Change Concept Generalizations
- Change is everywhere.
- Change is related to time.
- Change can be natural or manmade.
- Change may be random or predictable.

### Scientific Investigation Skills and Processes
- Make observations.
- Learn more.

### Assessment "Look Fors"
- Students can give examples of things that change.
- Students can categorize examples of change, explaining their reasoning.
- Students show understanding of the concept of "generalization."
- Students can give an example of a change generalization.

### Materials/Resources/Equipment
- Chart of Handout 2A (Five Senses Chart)
- Copies of Handout 2B (Change Is Everywhere), one for each student
- Three or four oranges
- One paper cup
- Drawing paper
- Four sentence strips with a different change generalization written on each strip
- Chart paper (to make Taba concept model chart)
- Student log books
- *My Five Senses* by Aliki (optional: use if background is needed about the five senses)

> **Note to Teacher:**
> - Some nonexamples of change include time, gravity, and the past.
> - Post a chart with the change generalizations in the classroom.

## Implementing the Lesson

1. Tell students they are going to explore a concept that is important to understanding our world—*change*. Explain that scientists spend a lot of time either trying to explain why a change occurred or causing a change to occur. Ask students how scientists study change.

   DOI: 10.4324/9781003238386-9

2. Explain that scientists often study change using their senses. Ask students to name the five senses. Post the Five Senses Chart (Handout 2A) in the classroom. Perform the following tasks to demonstrate the different ways that change can be observed and record student observations on the chart:
   - Show students some whole oranges. Ask students to describe the oranges only by looking at them.
   - Peel the orange. Ask students to describe how the orange changed. Ask students:
     - What sense first detected the change?
   - Pass around a few orange slices. Ask students to smell a slice and describe the smell of the orange.
   - Have the students close their eyes. Squeeze some orange slices so the cup catches the juice. Ask students:
     - What kind of change did you *hear*?
     - How did the orange change?
   - Have a volunteer *taste* the orange juice and describe it to the class. Ask students:
     - Could you make the taste of the orange juice change?
     - How could you make the taste change?
   - Let each student *feel* the orange slices that have had the juice squeezed out. Have them feel and describe the change.

3. Caution students that scientists have to be careful when using certain senses to determine change because they could be harmed. Ask students to provide examples of how using a certain sense to determine a change might be harmful.
4. Display a Taba concept model chart such as the one in Figure 2.

| Examples of Change | Categories of Change |
|---|---|
| Nonexamples of Change | Generalizations About Change |

**Figure 2.** Sample Taba concept model chart.

5. Ask students to draw an example of change and to share their drawings at their table.
6. Discuss what it means to classify things. Model the classification of objects into different categories. Remind students that in order to classify objects, they must find some way in which the objects are similar. Suggest that one category might be *people*. Ask students to bring their drawings to the front of the room if they matched the *people* category. Tape up the examples. Ask students to think of another category and continue the process of taping up the categories until all of the drawn examples of change have been taped.
7. Ask students to think about things that do not change. Can they come up with an idea? If this is difficult, suggest that they ask their parents this evening to see if they can think of something that does not change.
8. Discuss as a class why it was easier to think of examples that do change.

9. Explain that scientists often generalize or make statements about how examples are similar, and these generalizations help scientists understand our world. Ask students what they know about change by looking at the examples and nonexamples. Provide the students one example as a model (e.g., "I notice that there are many different kinds of changes" or "Everything changes"). Write down your statement on the section of the class chart labeled "Generalizations About Change."

10. Present the following generalizations by posting sentence strips in the section of the class chart labeled "Generalizations About Change" and explain to students that they will be looking at how these generalizations help scientists understand changes. Ask students to talk about each one and give some examples.
   - Change is everywhere.
   - Change is related to time.
   - Change can be natural or manmade.
   - Change may be random or predictable.

11. Discuss with students how the orange example applies to the change generalizations. Model by saying that the orange example showed that change is related to time (e.g., "I noticed some time went by in order for the orange to change."). Question the children in order for them to relate the generalizations to the orange experiment (e.g., "The change was manmade because you squeezed the juice out of it."). Ask the students if they can think of a natural change to an orange if it was left in a grove.

## Concluding and Extending the Lesson

### Concluding Questions and/or Actions
   - Have students complete one of the following statements orally:
      o  I like change because . . .
      o  I do not like change because . . .

   - How did your senses help you learn about change?
   - If you were a scientist, what changes would you like to study?

### What to Do at Home
   - Ask the students to work on the Change Is Everywhere handout (Handout 2B) with an adult family member.

Name: _____

Date: _____

# Five Senses Chart

| | | | | |
|---|---|---|---|---|
| **Sight** | | | | |
| **Sound** | | | | |
| **Smell** | | | | |
| **Touch** | | | | |
| **Taste** | | | | |

Name: _____

Date: _____

# Change Is Everywhere

Draw and label four things that change in your house.

```
┌──────────────┐        ┌──────────────┐
│              │        │              │
│              │        │              │
│              │        │              │
│              │        │              │
└──────────────┘        └──────────────┘

          ⬤ Change Is Everywhere ⬤

┌──────────────┐        ┌──────────────┐
│              │        │              │
│              │        │              │
│              │        │              │
│              │        │              │
└──────────────┘        └──────────────┘
```

# Lesson 3:
# What Scientists Do—Observe, Question, Learn More

## Planning the Lesson

**Note to Teacher:**
Gather two types of flowers for each group of 3 or 4 students. Each group should have the same two types of flowers.

### Instructional Purpose
- To apply three of six investigation processes (make observations, ask questions, and learn more) described in the Wheel of Scientific Investigation and Reasoning.
- To initiate an investigation of flowers.

### Instructional Time
- 45 minutes

### Change Concept Generalization
- Change is everywhere.

### Scientific Investigation Skills and Processes
- Make observations.
- Ask questions.
- Learn more.

### Assessment "Look Fors"
- Students can apply the scientific investigation process.

### Materials/Resources/Equipment
- Lab coat for teacher
- One lab coat (white adult T-shirt or dress shirt) for each student
- Chart, PowerPoint slide, or transparency of Handout 1B
- Chart, PowerPoint slide, or transparency of Handout 3A (Observations of Two Flowers)
- Two types of flowers for each pair of students
- Log book page for each student using Handout 3A or drawing paper for each student
- Chart paper
- Student log books

## Implementing the Lesson

1. Put on lab coats. Explain to students that they are going to learn to "think and work as a scientist" using the processes of scientific investigation.
2. Ask students to share their responses from the homework assigned during the previous lesson (Handout 2B):
   - What would you investigate/do if you were a scientist?

3. Ask students what they think it means when we say that scientists "investigate" something. Explain that to investigate something means you find out as much as you can about it. Ask students:
   - Have you ever investigated something?
   - What did you investigate?
   - How did you investigate?

4. Use the Wheel of Scientific Investigation and Reasoning (Handout 1B) to review the six processes introduced in Lesson 1 and explain that scientists do these things to "investigate" or learn about something: (1) make observations, (2) ask questions, (3) learn more about observations and questions, (4) design and conduct experiments, (5) create meaning, and (6) tell others what was found. Remind students that scientists use these processes when learning about their world.

5. Point to the Make Observations section on the wheel. Tell students that the first thing scientists do is use their senses to learn about something. Describe the senses and the body part associated with each of the senses.

6. Explain to the class that they are going to use their senses to observe flowers. Divide the students into pairs and give each group two different kinds of flowers. Pick up one type of flower and ask each group to make observations of the flower they have as you write their observations on the chart of Handout 3A. Ask students:
   - What do you notice about the flower? What do you observe about the smell, the color, the petals, and the leaves?
   - When you make observations, you use your senses to learn. What sense do you use most to make observations?
   - What senses did you use to make these observations?

7. Ask each group to examine the second flower and make observations. Write their observations in the second column on the chart of Handout 3A.

8. Point to the Ask Questions section on the wheel. Have students think about questions they could ask when observing the two flowers. Ask students:
   - What questions do you have about flowers?

   If students are hesitant, the teacher should give an example of a question she or he has to encourage the students. Help students understand the difference between making a statement and asking a question. If students make a statement, rephrase the statement as a question. Write the students' questions on chart paper. Then model your question, recording it on the board, PowerPoint slide, overhead, or chart paper. Some questions might include:
   - How might you learn if one flower has more petals than another?
   - How are the flower petals alike and how are they different?
   - Why do you think that flowers have petals?

9. Point to the Learn More section on the wheel. Ask students how they can gather information to learn more about a topic. Encourage them to think of the many ways people learn. Emphasize that the more they observe something, the more they can learn about it. Ask students what they noticed about the differences in the flowers. Use the following questions:
   - How are the flowers alike? How are they different?
   - What questions do you have about flowers that are not already on our list?
   - When you want to learn more about something, what do you do?

10. Review the scientific investigation skills and processes that you covered today: Make Observations, Ask Questions, and Learn More.
11. Tell students that they are going to learn more about plants and animals through investigation. Distribute a copy of Handout 3A to each student and ask the students to either draw or write about their observations of the two flowers. Staple the completed worksheet into student log books.

## Concluding and Extending the Lesson

### Concluding Questions and/or Actions
- Share log entries about what students observed.
- Provide props and encourage students to try making observations, asking questions, and learning more.
- Provide copies of books about flowers and plants in your classroom library.
- Provide different kinds of flowers in the science investigation center.

### What to Do at Home
- Ask the students to observe other types of flowers that they may find in their neighborhood, on television, on the Internet, in books and magazines, or in gardening stores.
- Encourage the students to share their observations of similarities and differences to the flowers observed in class.

Name: _____  Date: _____

# Observations of Two Flowers

| Sense | Flower #1 | Flower #2 |
|-------|-----------|-----------|
| Smell |           |           |
| Touch |           |           |
| Sight |           |           |

# Lesson 4:
# What Scientists Do—Experiment, Create Meaning, Tell Others

## Planning the Lesson

### Instructional Purpose
- To design and conduct an experiment using flowers.
- To create meaning from the experiment.
- To learn how to tell others what was found.

### Instructional Time
- 45 minutes

### Change Concept Generalizations
- Change is everywhere.
- Change is related to time.
- Change can be natural or manmade.
- Change may be random or predictable.

### Scientific Investigation Skills and Processes
- Make observations.
- Ask questions.
- Learn more.
- Design and conduct the experiment.
- Create meaning.
- Tell others what was found.

### Assessment "Look Fors"
- Students can design and conduct an experiment.
- Students can interpret data from a data table.
- Students can describe how the experiment was conducted and what results were found.

### Materials/Resources/Equipment
- Lab coat for teacher
- One lab coat (white adult T-shirt or dress shirt) for each student
- Chart of Handout 1B
- Charts, PowerPoint slides, or transparencies of Handouts 4A (Definition of Hypothesis), 4B (Steps for Flower Experiment), and 4C (Flower Experiment Data Table)
- Copies of Handout 4C, one for each student
- One badge per student using Handout 4D
- Two types of flowers with a few clearly defined petals (one flower of each type per pair of students)
- Chart paper
- Pencils
- Student log books

> **Note to Teacher:**
> Gather two types of flowers for each group of 3 or 4 students. Each group should have the same two types of flowers.

## Implementing the Lesson

1. Put on lab coats and ask students to share their findings from the homework assignment, using the following questions:
   - What flowers did you observe outside of school?
   - What did you notice about the flowers?
   - What might cause flowers to change?

2. Review what the class did during the previous lesson (Lesson 3), by asking:
   - What did we start investigating in our last lesson?
   - What did we do to begin our investigation of flowers?
   - What did we observe about the flowers?
   - What questions did we have about the flowers?
   - What question did we decide to investigate?

3. Tell students: "In the last lesson, we investigated flowers to practice making observations, asking questions, and learning more. Today, we will continue thinking as scientists, and we will design and conduct an experiment with flowers. This unit, *Survive and Thrive,* is about both plants and animals. We will conduct an experiment with insects later in the unit." Record student responses to the following questions on chart paper:
   - What observations can you make about the flowers?
   - What questions do you have about flowers?
   - What can we do to learn more?

4. Point to the Design and Conduct the Experiment section on the wheel. Note that the first thing scientists do to conduct an experiment is to form a hypothesis from their question or questions. Use Handout 4A to define hypothesis as "a temporary prediction that can be tested about how a scientific investigation or experiment will turn out" (Scholastic, 2007). Explain to the students that they are going to design and conduct an experiment to answer the question, "Do the two types of flowers have the same number of petals?" Tell students that they will be forming a hypothesis based on this question.

5. Explain that the hypothesis needs to be tested. Ask students what they think to "test a hypothesis" means. Explain the definition of a hypothesis according to the glossary. Develop a general hypothesis. Then encourage students to share how they think the class could find out whether the prediction or hypothesis is true.

6. Explain that it is important to plan the experiment by listing the steps. Ask students what steps they would take. After students share, reveal the list of steps the class is going to follow (see Handout 4B). Point out the list of materials that are needed for the experiment.

7. Post a data table that is based on Handout 4C. Using the PowerPoint slide, transparency, or chart of Handout 4C, show students the question at the top of the page: "Do the two types of flowers have the same number of petals?" Then have each group of students decide on a prediction.

8. Explain that the class will work as scientists to conduct the experiment and collect the data. They will look at the data and think about what it means. Finally, they will answer the question, "Do the two types of flowers have the same number of petals?"

9. The teacher should model how carefully the petals should be pulled off one flower and counted. Have the groups of students pull off the petals, one at a

time, and count them. Record each group's number in the data chart. Then the small groups should do the same thing with the second flower. (The teacher may lead the group and everyone can do it together if preferred.)

10. After all of the data have been recorded, ask students to think about what the data mean and answer the question, "Do the two types of flowers have the same number of petals?" By doing this, explain that this process helps them Create Meaning.

11. Tell students that they have just conducted a scientific investigation or experiment. They tested their hypothesis and now they need to tell others what they found out. Point to the final section on the wheel: Tell Others What Was Found.

12. Ask students how they could share this new information with other people. Allow students to think about and talk about ways to do this. Tell students that they could show others the data table they made, they could tell people, or they could draw pictures.

13. Proclaim that the student scientists have just conducted an experiment and pass out "badges" saying "I Conducted an Experiment in Science—Ask Me About It" (Handout 4D). Also ask students to date and make the following entry in their student log books:
    - Draw a picture of your group conducting the flower experiment we did today when thinking as scientists.

## Concluding and Extending the Lesson

### Concluding Questions and/or Actions
- Share student log book entries.
- Explain how we worked as scientists today. What were the steps we did for the experiment?
- Why do you think scientists do experiments?
- Did you notice any changes in our data based on the type of flower and number of petals?
- How do you think the number of petals on the two flowers might change over time?
- Do you think other types of flowers might have different numbers of flowers? How could you find out?
- Have different kinds of flowers available in a scientific investigation center so that students can investigate.

### What to Do at Home
- Encourage the students to tell their parents or an adult family member about the steps of a scientific experiment.
- Encourage the students to ask their family member whether there is an experiment that they could do as a family.

# Definition of Hypothesis

## A hypothesis is . . .

"a temporary prediction that can be tested about how a scientific investigation or experiment will turn out."

(Scholastic, 2007)

Name:_____ Date:_____

# Steps for Flower Experiment

**Hypothesis:** We think the two types of flowers will have (circle one):

The SAME number of petals.

A DIFFERENT number of petals.

## Experiment Steps:
1. Take one flower and count the number of petals on the flower.
2. Write the number in the correct space on the data table.
3. Take the second flower and count the number of petals on the flower.
4. Write the number of petals in the correct space on the data table.
5. Think about what the data table tells us.
6. Answer the question, "Did the two types of flowers have the same number of petals?"

## Materials Needed:
- Flowers
- Pencils
- Handout 4C

Name:_____ Date:_____

# Flower Experiment Data Table

**Question:** Do the two types of flowers have the same number of petals?

**Hypothesis:** _____

| Group | Flower #1 Number of Petals | Flower #2 Number of Petals |
|-------|----------------------------|----------------------------|
| #1 | | |
| #2 | | |
| #3 | | |
| #4 | | |
| #5 | | |

**Findings:**

# Science Investigation Badges

I Conducted an Experiment in Science— Ask Me About It!

I Conducted an Experiment in Science— Ask Me About It!

I Conducted an Experiment in Science— Ask Me About It!

I Conducted an Experiment in Science— Ask Me About It!

I Conducted an Experiment in Science— Ask Me About It!

I Conducted an Experiment in Science— Ask Me About It!

# Lesson 5:
# What Is a Life Cycle?

## Planning the Lesson

### Instructional Purpose
- To demonstrate an understanding that all plants and animals undergo different changes as they grow and develop.

### Instructional Time
- 45 minutes

### Change Concept Generalizations
- Change is everywhere.
- Change is related to time.
- Change may be random or predictable.

### Key Science Concepts
- All plants and animals undergo different changes in their life cycles.
- As animals and plants grow; they get larger according to a pattern.
- Animals are similar to their parents.

### Scientific Investigation Skills and Processes
- Make observations.
- Ask questions.
- Learn more.

### Assessment "Look Fors"
- Students can express the idea that living things grow and change.
- Students can explain the life cycle of living things.

### Materials/Resources/Equipment
- One copy of Handout 12B (Scientific Investigator Certificate)
- Charts, PowerPoint slides, or transparencies of Handouts 5A (Life Cycle of a Plant) and 5B (Life Cycle of a Frog)
- Copies of Handouts 5B and 5C (Personal Life Cycle), one for each student
- Scissors
- Glue
- Chart paper
- Markers
- Construction paper

> **Note to Teacher:**
> The picture-sequencing activity will indicate whether students understand the concept of things growing and changing. If students get the pictures in the wrong order, engage them in conversation to see that they change the order and are able to articulate why they made the changes.

## Implementing the Lesson

1. Remind students that they are thinking as scientists. Explain to students that they are working toward earning a Scientific Investigator Certificate (Handout 12B). The certificate is earned by participating in the lessons, practicing

 DOI: 10.4324/9781003238386-12

thinking as a scientist, and completing assignments in their log books. Let students know at the end of each lesson if they have met the criteria for lesson participation. Today's lesson is the first step in the journey toward earning their certificate. Tell them that today's lesson is about survival.

2. Explain that all living things go through a life cycle. Tell students that, starting with the next lesson, they will have the opportunity to examine the life cycle of a living organism.

3. Ask students what they think life cycle means. Have them define "life" (e.g., living, from birth to death) and "cycle" (e.g., a circle, stages that continue).

4. Begin a discussion about life cycles by asking the following questions:
   - What were you like when you were a baby?
   - How have you changed?
   - Did this change happen quickly or slowly?
   - How are your family members alike? How are they different?
   - Have you seen your family members change over time? If so, how have they changed?

5. Show students the PowerPoint, transparency, or chart of Handout 5A. Explain the diagram to them, pointing out the stages of the plant life cycle.

6. Continue the life cycle discussion you started with the students earlier, asking:
   - What do you think the life cycle of an animal would be like?

7. Give students Handout 5B. Ask students to look at the pictures and describe what they see. Tell them that they are going to make a model to illustrate the life cycle of an animal. Instruct them to cut out the pictures and put them in chronological or sequential order. As students are working, engage them in informal conversation about their thinking.
   - Why did you put this picture first?
   - What did you think about to help you decide which one came first?
   - Look at the pictures the other students put in order. Do all of you have your pictures in the same order?
   - Do you think you need to change anything on your paper? Why or why not?
   - What picture did you put second? Why?
   - What did you think about to help you decide what came second?
   - Can you think of an explanation for everyone having their pictures in the same order?
   - If we were going to add another picture to show what happens next, what would it look like? What story do our pictures tell?
   - What do our pictures show us about life cycles?

Once the teacher has confirmed that the pictures are in the correct order, students should glue them on construction paper.

## Concluding and Extending the Lesson

### Concluding Questions and/or Actions
- Which concept generalizations about change apply to our study of life cycles?
- What can we infer about living things and life cycles?
- Why is it important to understand life cycles?
- Do all living things go through life cycles? Explain your answer?
- Can we skip a step in a life cycle?

### What to Do at Home

- Ask the students to find at least three pictures of themselves or another family member (or draw them) and put them in chronological order (Handout 5C).
- Encourage the students to bring the pictures to class and describe what changes occurred between each picture.

# Life Cycle of a Plant

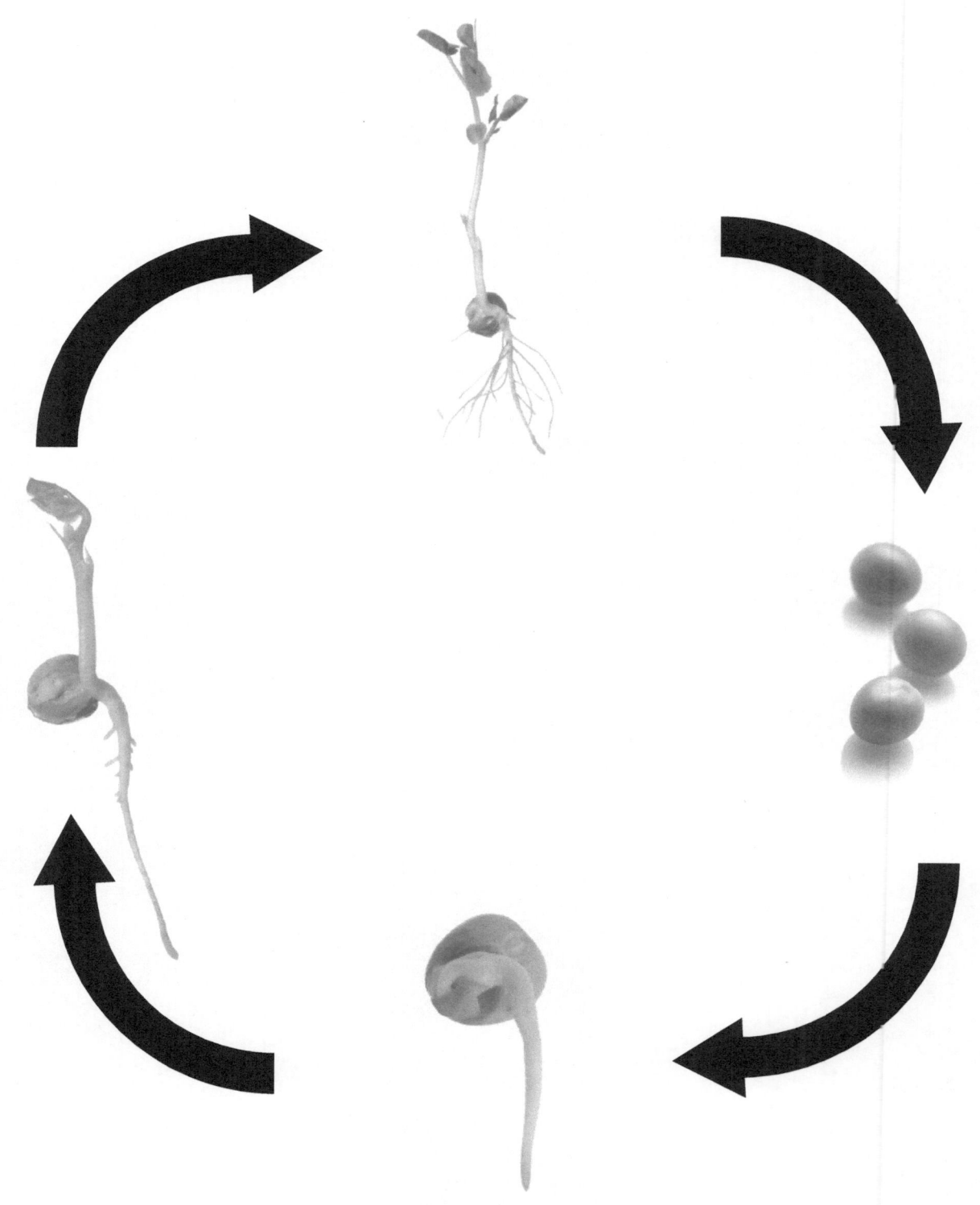

# Life Cycle of a Frog

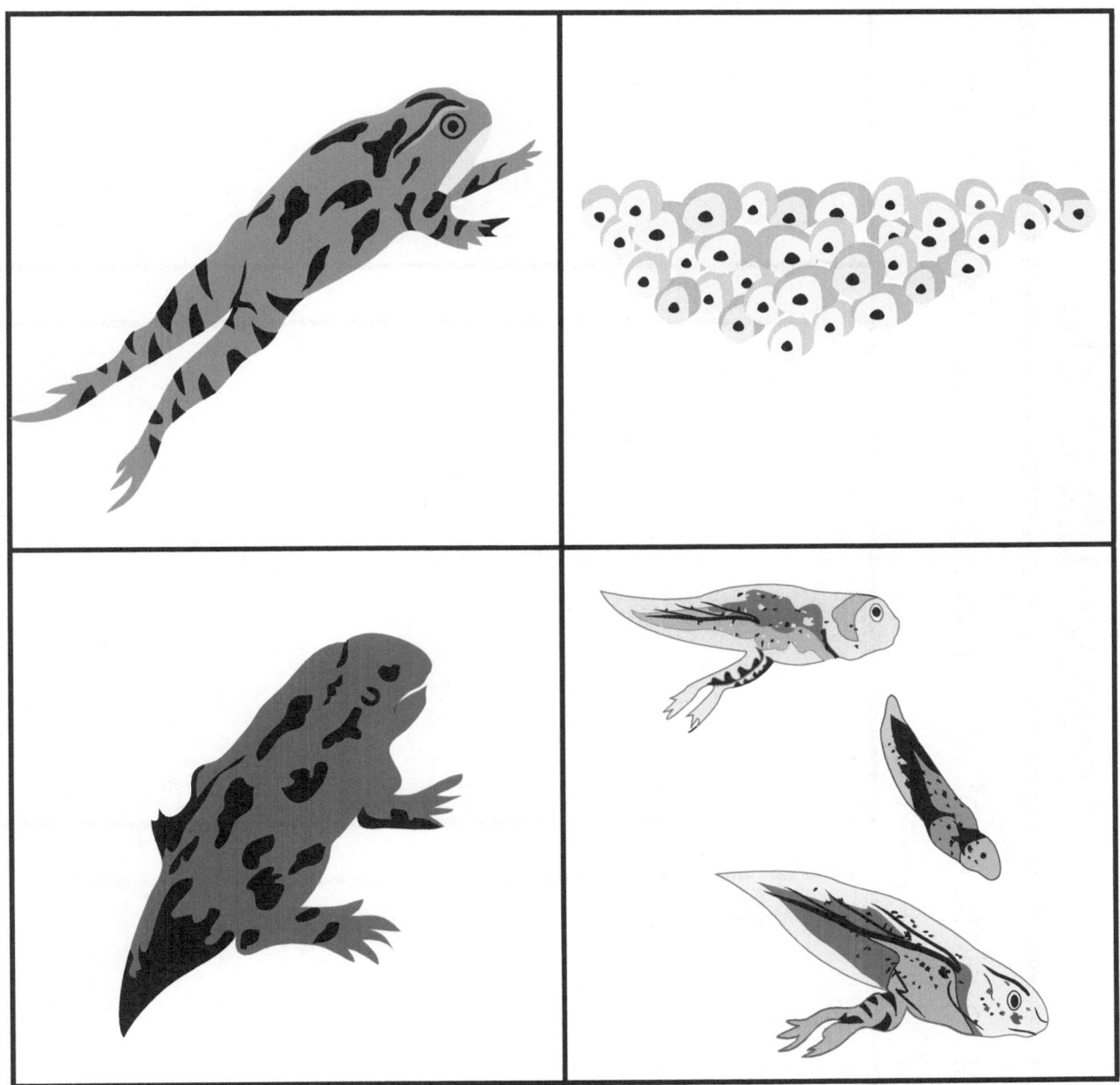

Name: _____ Date: _____

# Personal Life Cycle

Find or draw three pictures of you or someone you know. Put them in order to show change and life cycle.

# Lesson 6:
# What Is the Life Cycle
# of a Mealworm?

## Planning the Lesson

### Instructional Purpose
- To investigate the life cycle of a mealworm.

### Instructional Time
- Two 45-minute sessions

### Change Concept Generalizations
- Change is related to time.
- Change can be natural or manmade.
- Change may be random or predictable.

### Key Science Concepts
- Plants and animals undergo different changes in their life cycles.
- Animals are similar to their parents.

### Scientific Investigation Skills and Processes
- Make observations.
- Ask questions.
- Learn more.
- Design and conduct the experiment.
- Create meaning.
- Tell others what was found.

### Assessment "Look Fors"
- Students can give reasons to support their hypotheses.
- Students can describe the life cycle of a mealworm.
- Students can record observations on a data table.
- Students can report findings on the Experimental Report Form.

### Materials/Resources/Equipment
- Lab coat for teacher
- One lab coat (white adult T-shirt or dress shirt) for each student
- Chart of Handout 1B
- Charts, PowerPoint slides, transparencies, or posters of Handouts 6A (Mealworm Investigation Hypothesis) and 6B (Steps for Mealworm Experiment)
- PowerPoint slide, transparency, or poster of Handout 6E (Completed Life Cycle of a Mealworm, to be used as a review following the experiment but before the assessment)
- Charts, PowerPoint slides, posters, or copies of Handouts 6C (Mealworm Data Table), 6D (Experimental Report Form), 6F (Life Cycle of a Mealworm), 6G (Living Things Concept Map), and 6H (Animals Observation Sheet), one for each student

 DOI: 10.4324/9781003238386-13

- Live mealworms (may be purchased online, at a pet store, or bait stores often in containers of 50)
- Clear plastic shoe box with loose-fitting lid
- Oatmeal, bran meal, or wheat meal
- Apples or potatoes
- Magnifying glasses
- Rulers
- *Mealworms* by Donna Schaffer
- Student log books

## Implementing the Lesson

### Session 1

1. Put on lab coats and remind students that they are thinking as scientists and working toward earning their Scientific Investigator Certificate. Today's lesson is the second step in the journey toward earning their certificate. Tell them that today's lesson is about survival and change.
2. Review the life cycles from yesterday's lesson. Review the pictures from the homework and talk about the human life cycle and about how the students or their family members have changed over time. Ask students:
   - Is a life cycle predictable or unpredictable?
   - Do students resemble their families?

3. Tell students that they will continue to think like scientists to determine the changes that take place in the life cycle of an insect.
4. Define "worm" (a soft, legless invertebrate animal that has a long, movable body) and "habitat" (the place where a plant or animal lives) on the board. Explain the words and use them as appropriate during the lesson.
5. Show the mealworms to the students. Explain that they will be doing an experiment with the mealworms over the next few weeks.
6. Give each student a mealworm to examine carefully. When choosing mealworms for the experiment, you should choose the largest ones, as they will be the most mature and closest to metamorphosis. Keeping the mealworms in a warm place also will encourage quicker metamorphosis. The students should use magnifying glasses to look at the mealworms.
7. Direct students to observe the different body parts, colors, and markings. Ask them to infer what kind of animal this is. Ask them to make a guess based on what they know (hypothesize).

**Note to Teacher:**

Over time, the mealworm will become less active. The mealworms will shed their exoskeletons. Children will find these "sheds" in the meal. They should be able to infer what these are. The mealworm will curl up, and, after its last larval molt, it will change into the pupa form. Usually, children will see one of the larvae as it is changing. Be sure to allow all students to observe it during this critical time.

The pupa stage will last from 10 to 20 days. Pupae do not eat or move. Warmth (room temperature) will aid the metamorphosis. After the pupa's hard shell splits, the adult mealworm (beetle) emerges. Again, children may be able to see the beetle emerging from the pupa. After the adult beetles emerge, they will mate and lay eggs on the bottom of the plastic box. Students will be able to see tiny circles of eggs surrounded by a sticky liquid to protect the eggs stuck to the plastic. The circles grow larger, and then tiny mealworms emerge from them in about 2 weeks. The newborns are very small. You will need to look very closely to see them. They will be the color of the meal.

It will take up to 90 days for a larva to grow to the pupa state. That is why we begin the experiment with the largest larva possible. An excellent book on the mealworm lifecycle is *Mealworms* by Donna Schaffer, published by Bridgestone Books, Mankato, MN. This book should be used with the students *after the experiment* has been finished; however, it is a good source for teachers who have no experience with mealworms.

If students infer that the mealworms are worms, refer them to the definition of "worm" on the board. Allow students to make hypotheses about what the mealworm is, what stage of development it might be in, and what type of animal it is. Is it like a caterpillar or something else?

8. Explain to students that they will be learning more about the basic needs (oxygen, food, water, and shelter or a place to grow) of plants and animals in later lessons. For mealworms, their food is meal (oatmeal, bran meal, or wheat meal). They get water from the potato or apple. Their shelter is the box where they live. Explain to students that it is important for the mealworms to have their needs met so that they can survive.

9. Use the Concept Mapping Section in Appendix B as a guide for instructing students in concept mapping. To begin this activity, tell students: "We have just finished studying about life cycles. Let's review what we have learned. I am going to teach you how you can show what you have learned in a way that will help you remember." Show them the PowerPoint slide or overhead of Handout 6G or distribute copies of Handout 6G. Tell students: "We know that animals are living things." Use the following questions as a guide to complete the concept map activity (see Figure 3 for a completed map). See the example below for guidance.

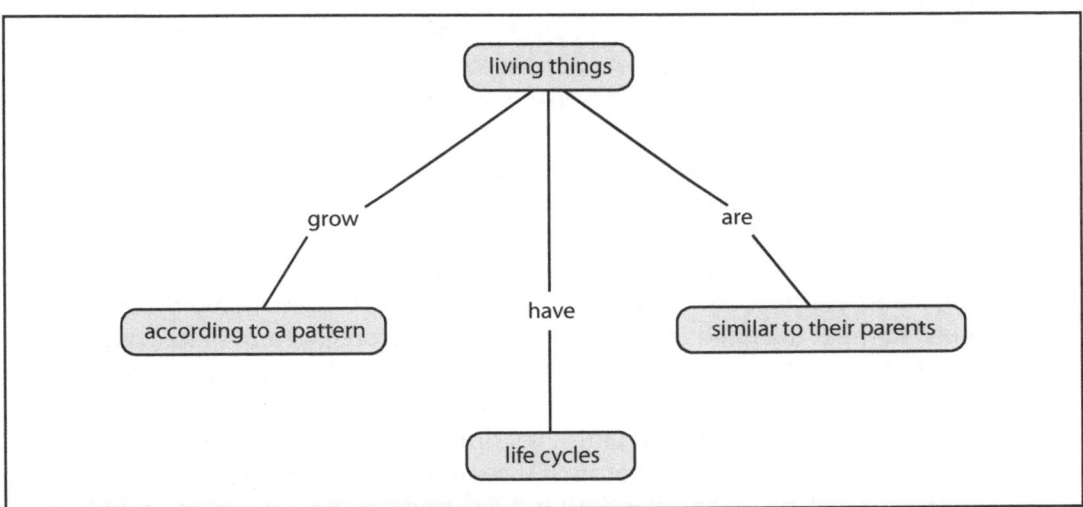

**Figure 3.** Completed concept map.

- What can we say about living things?
- What did we learn about the growth of living things? (Remind them about the plant and frog examples from Lesson 5.)
- What can we say about the children or offspring of living things?

10. Review what the class discussed about life cycles from the previous lesson. Engage students in a discussion using the following question as a guide:
   - What do you think you will find out about the life cycle of a mealworm? Why do you think this?

Explain that students are going to observe the mealworms for several weeks to see what they can find out about the life cycle of a mealworm.

## Session 2

11. Put on lab coats and review Handout 1B (the Wheel of Scientific Investigation and Reasoning). Ask students:
    - What do scientists do?
    - What skills do scientists have?
    - What steps do scientists go through to conduct an experiment?

12. Display a poster of Handout 6A. Explain that scientists plan their investigations before they begin and write down their observations as the experiment goes on.

13. Ask students what kinds of questions they might be asking about mealworms. What kind of investigation might we be conducting? (Try to probe until they come up with: "What kind of life cycle does a mealworm have?")

14. The first thing scientists must do is generate a prediction or hypothesis. Ask students to hypothesize what they think they will find out. Will the mealworms have live babies? Lay eggs? Grow into snakes? Phrase the hypothesis as "I think the life cycle of a mealworm is . . ."

15. Have students explain their hypotheses, giving reasons to support them. Write these hypotheses on the board or chart paper. Students may write their own hypotheses in their log books.

16. Show students the materials that will be available for their use: clear plastic box with lid; ruler; oatmeal, wheat meal, or bran meal; apple or potato; magnifiers. Teachers may elect to have one large container to house all mealworms in the class or to provide students with smaller containers so that they may each have a few mealworms of their own.

17. Discuss what will be needed for the investigation and what will not be needed.

18. Write a list on the board or on chart paper.

19. Brainstorm with students to generate a list of the steps necessary to complete the investigation. Communicate safety and care issues, including that they will need to meet the needs of the mealworms (e.g., food, water, shelter, oxygen).

20. To learn about the mealworm's life cycle, students will need to keep the mealworms for a period of time. Provide students with multiple copies of Handout 6C to help them organize the drawings of their observations and their notes about the mealworms. (Depending on students' ability level, tell them they will be given a handout in a few minutes. If students are unable to complete this on their own, create a chart to be completed by the class as a whole.)

21. Steps for completing the experiment are included in Handout 6B. The steps should include the creation of a habitat for the mealworms, a feeding/ watering schedule for the mealworms, and an observation schedule. Students will also need to create a chart or table for recording the data they collect in their logs. Show them the tables from earlier lessons and see what they are able to develop. Teachers may need to assist in this process, or this can be done as a whole class. Write the steps on the board or chart paper or create a transparency, slide, or chart of Handout 6B. Explain to students that they will begin following the steps to make a mealworm habitat and will begin making observations. Distribute copies of Handout 6C and explain it to the students.

22. Have students respond to the following prompt in their log books:
    - Today I thought like a scientist by . . .

## Concluding and Extending the Lesson

### Concluding Questions and/or Actions
- What is the problem or question we are investigating?
- What is your hypothesis?
- What steps will you take to continue thinking as a scientist?

### What to Do at Home
- Ask students to make observations of a pet using Handout 6H. If students do not have a pet, suggest that they ask a friend or family member who might have a pet if they could observe their pet.

### Continuing Observations
- During subsequent observations, the meal should be spread out on a table or desk. Students should look through the meal and find the mealworms. They also will find "sheds" and "feces" in the meal. When questions arise about what students are finding, make sure they understand that all animals have waste products similar to these. Students also may find dead mealworms. However, be sure that these are not simply pupae in the process of metamorphosis. These should be observed and the observations recorded. Students also can draw pictures to compare the live and dead mealworms. After adult beetles have emerged, begin watching for the eggs on the bottom of the plastic box. Point these out to the students when they become evident. At the end of the experiment, students should complete the Experimental Report Form (Handout 6D) and Lifecycle of a Mealworm (Handout 6F). This may be done individually, in small groups, or as a whole group activity led by the teacher, depending upon ability level.

> **Note to Teacher:**
> Use the PowerPoint slide, transparency, or chart of Handout 6E to review the mealworm life cycle with the students prior to giving them the assessment (Handout 6F).

Name:_____ Date:_____

# The Mealworm Investigation
# Hypothesis

**Question:** What kind of life cycle does a mealworm have?


**Hypothesis:** I think

_____

_____

_____

_____

_____

# Steps for Mealworm Experiment

1. Gather materials: box with lid, oatmeal, potato, mealworms.

2. Put oatmeal into the box until it is one inch deep.

3. Put a slice of potato into the box.

4. Count the live mealworms and record the date and the number.

5. Place the mealworms into the box and put the lid on it.

6. Each day or two, observe the mealworms (with and without magnifiers) and record what you observe.

7. Keep a chart showing how many mealworms are counted each day and record any additional information or observations.

8. Remember to wash your hands.

Name:_____ Date:_____

# Mealworm Data Table

| Observation | Draw Your Mealworm | Changes |
|---|---|---|
| Observation #____<br><br>Date: _____ | | |
| Observation #____<br><br>Date: _____ | | |
| Observation #____<br><br>Date: _____ | | |
| Observation #____<br><br>Date: _____ | | |
| Observation #____<br><br>Date: _____ | | |

Name:_____ Date:_____

# Experimental Report Form

Name of Experiment_____

1. What was your hypothesis (or prediction) about what would happen?

   _____

   _____

2. What materials did you use to test the hypothesis?

   _____

   _____

3. What methods did you use? (Outline the *steps* you took.)

   _____

   _____

   _____

   _____

4. What data have been collected? Where are your data recorded? (Attach your data table.)

_____

_____

5. What are your findings? (Did your hypothesis prove to be true or false?)

_____

_____

6. What new questions do you have?

_____

_____

_____

_____

## Handout 6E
# Life Cycle of a Mealworm

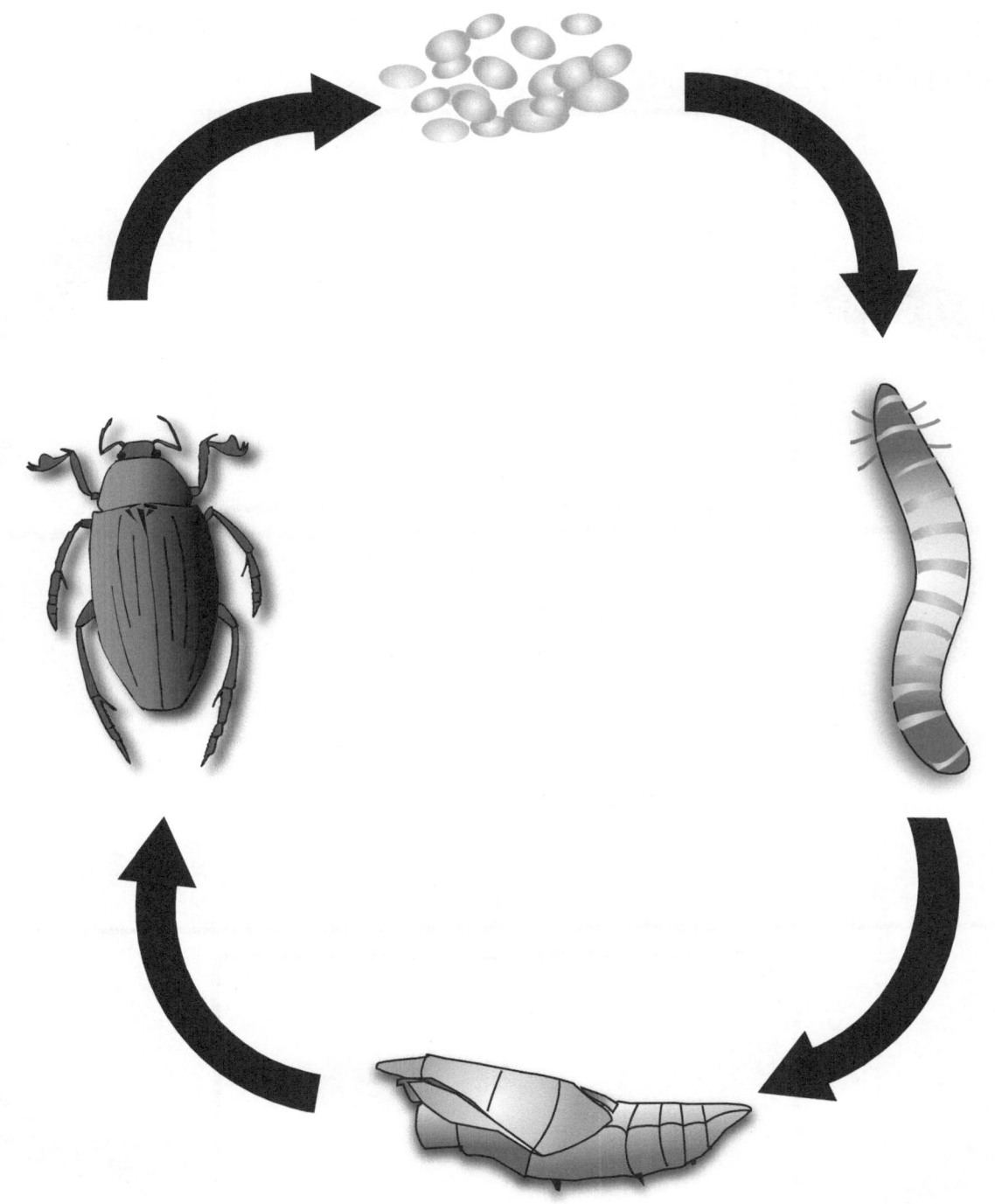

## Handout 6F
# Life Cycle of a Mealworm

Draw pictures to show the four stages of the life cycle of a mealworm.

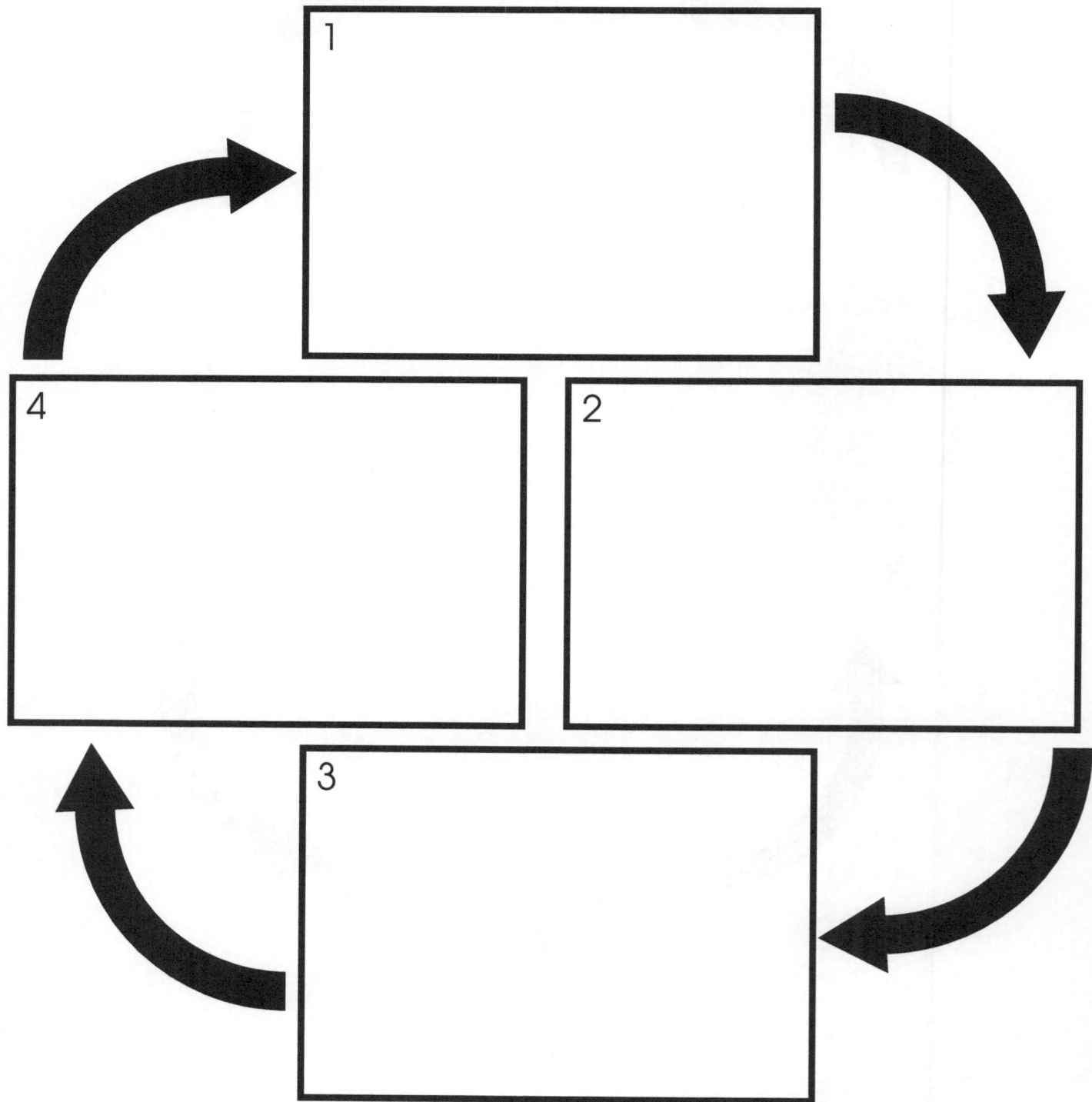

Name: _____                     Date: _____

# Living Things Concept Map

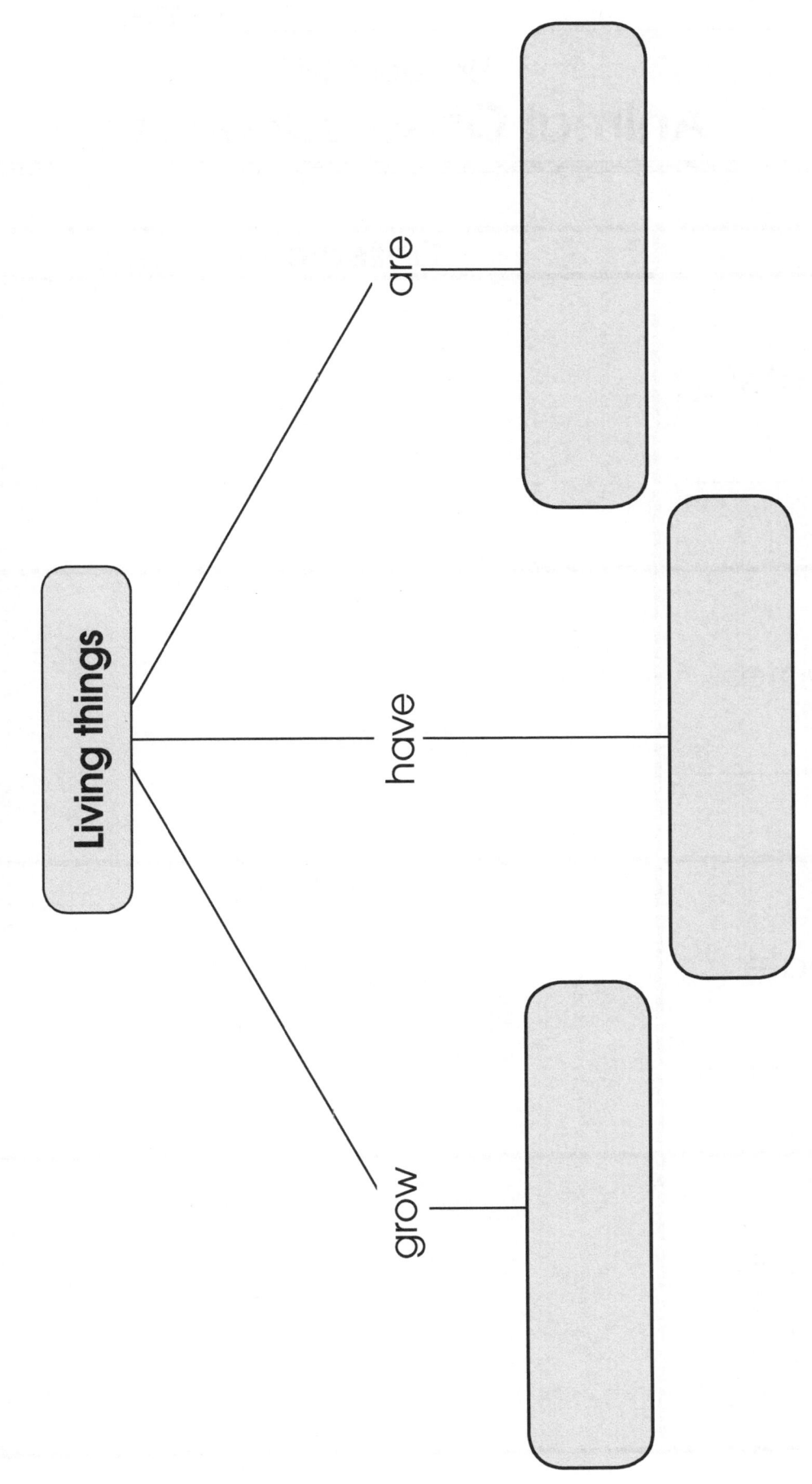

Name:_____ Date:_____

# Animal Observation Log

| | Observation Notes |
|---|---|
| Date:<br><br>_____ | |
| Date:<br><br>_____ | |
| Date:<br><br>_____ | |
| Date:<br><br>_____ | |

# Lesson 7:
# What Are the Requirements for Life?

## Planning the Lesson

### Instructional Purpose
* To identify the basic needs of all animals: oxygen, food, water, and shelter or a place to grow.

### Instructional Time
* 45 minutes

### Change Concept Generalizations
* Change can be natural or manmade.

### Key Science Concepts
* Plants and animals have basic needs for oxygen, food, and water.
* Animals need a suitable place to live.

### Scientific Investigation Skills and Processes
* Make observations.
* Ask questions.
* Learn more.

### Assessment "Look Fors"
* Students can state the basic needs of animals.

### Materials/Resources/Equipment
* Lab coat for teacher
* One lab coat (white adult T-shirt or dress shirt) for each student
* PowerPoint slide or transparency of Handout 6E (Complete Life Cycle of a Mealworm) or another picture of a worm
* PowerPoint slide or transparency of Handout 5B (Life Cycle of a Frog) or another picture of a frog
* PowerPoint slides and handouts of Handouts 7A (Blank Requirements for Life Concept Map)
* Several pieces of chart paper; one piece divided into a T with the headings: Needs/Wants
* Cup of water
* Food items (basic needs and wants)
* A few toys
* Variety of animal books
* Animal Flash Cards (http://www.trendenterprises.com) or a variety of animal pictures
* Student log books

 DOI: 10.4324/9781003238386-14

## Implementing the Lesson

1. Put on lab coats and remind students that they are thinking like scientists and working toward earning their Scientific Investigator Certificate. Today's lesson is the third step in the journey toward earning their certificate.

2. Place some items on a table in front of the students and ask them to gather around. These items should include water, food, and a couple of toys. On a piece of chart paper, divided into a T-chart labeled "Needs/Wants," ask students to tell you which items on the table are needs and which are wants. Correct any misconceptions. Generate a discussion with the following questions:
   - What is the difference between something you need and something you want?
   - What do we need to survive?
   - Do needs ever change? How?
   - Do wants ever change? How?

3. The *Scholastic Children's Dictionary* (2007) defines *survive* as "to continue to live or exist." List the student responses to "What do we need to survive?" on a piece of chart paper. Make certain that the list includes food, water, oxygen, and shelter or a place to grow. (If students do not come up with the idea that we need oxygen, show them a picture of an astronaut or a scuba diver and ask them about the equipment the astronaut or diver is wearing.) Ask students:
   - What happens when we don't have one or more of our needs met?

4. Divide students into pairs and ask students to describe their homes to each other. You might want to give them an example to get them started. Typical responses include that each home has a place to prepare food and eat it, water to drink, and a place to sleep. Some prompting questions include:
   - What do the homes have in common?
   - How are they different?
   - Do they differ in what's necessary to meet our needs?
   - Do we ever change our shelters? Why?

5. Explain to students that animals, like us, need oxygen, food, water, and shelter or a place to grow. Still, there are some differences. Show students Handout 6E or another picture of a worm and ask them if they would like to eat it. Then ask them which animal(s) would like to eat it. Next, show students Handout 5B or another picture of a frog. Go through the same process, asking them which animal(s) like to eat frogs. Chart student responses. Then, ask students the following questions:
   - Do animals need water? Oxygen?
   - What about shelter? Do some animals have houses like we do? In what kinds of places do animals live?

6. Show students photographs of an ecosystem. Generate a discussion with the following questions:
   - What basic needs do animals have?
   - Which animals might live in the river?
   - Which animals might live on the riverbanks?
   - How does living by water help fulfill basic needs?

Select one of the animals listed by the students and make sure that all of its needs are met.

7. Tell students: "We are going to use what you have learned today to make a new concept map." Display Handout 7A as a PowerPoint slide or a transparency or distribute copies of Handout 7A to the students. Use the following questions to guide the completion of the map. Provide time for students to complete the concept map.
   - Animals change and grow. Based on this information, what can we hypothesize about animals?
   - As animals grow and change, they require or need certain things to survive and thrive. What do they need?

8. Tell students that they are going to be making observations, learning more, and asking questions. Depending on their ability levels, give each student, pair of students, or group of 3 or 4 students a picture of an animal. (Animal flashcards or references from the resource list can be used here.) Have the students draw or write about their animal's oxygen, food, water, and shelter needs in their log books. Students should find out what their animal eats and where it lives. They should use prior knowledge as well as the books that are available to research their animals. Students should be able to learn a lot of the information just from the pictures that are in the books. Tell students as they work also to think of a new question they now have about animals. At the bottom of their drawing or writing, students should write a new question (or orally tell the teacher the question and have it written down) that they now have about their animal.

9. Tell students: "Let's relate this to what we have learned about change. Animals live in different kinds of shelters based on their needs. For example, if a bird is living in a nest in a tree and the tree is chopped down, what will the bird do?" (Possible responses might include: Find a new habitat such as a tree or gutter to live in, migrate, or die.) Ask:
   - Which change generalization(s) does this fit? (1) Change is everywhere. (2) Change is related to time. (3) Change can be natural or manmade. (4) Change may be random or predictable.
   - Can you think of other examples in nature?

## Concluding and Extending the Lesson

### Concluding Questions and/or Actions
- Have students respond to the following prompt in their log books:
  o "In order to survive, animals need . . ."

### What to Do at Home
- Ask students to continue their pet observations.
- Ask the students to discuss their needs and wants with their parents or other adult family member. Ask the students to ask questions such as, "Are all foods needs? How do you decide what is a "need" and what is a "want?"

### Handout 7A
# Blank Requirements for
# Life Concept Map

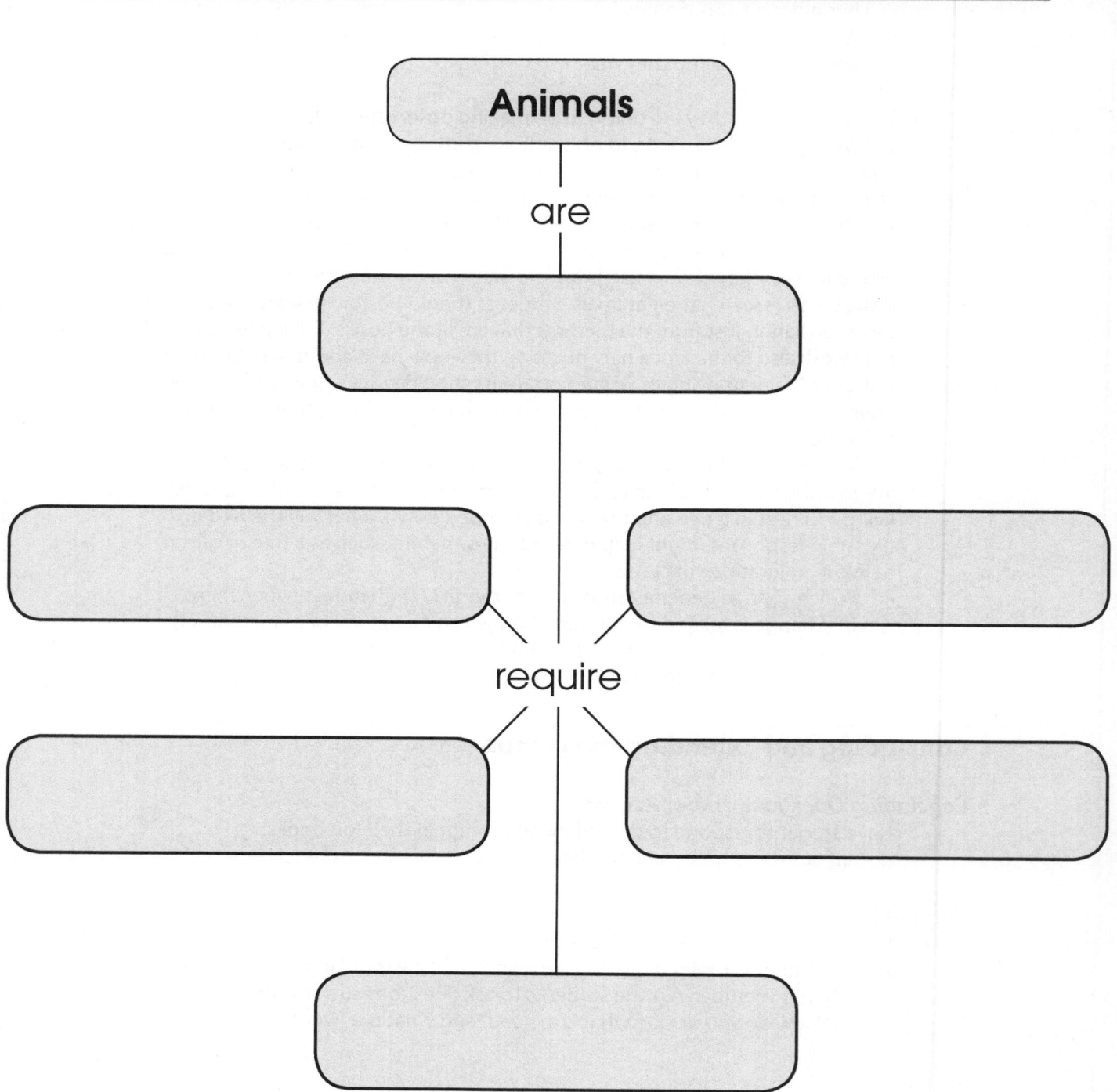

Animals

are

require

# Lesson 8:
# What Do Plants Need to Thrive?

## Planning the Lesson

### Instructional Purpose
- To understand the needs of plants.
- To design an experiment using the Wheel of Scientific Investigation and Reasoning.
- To understand the difference between surviving and thriving in plants.

### Instructional Time
- 45 minutes

### Change Concept Generalizations
- Change is related to time.
- Change can be natural or manmade.

### Key Science Concepts
- Plants and animals have basic needs for oxygen, food, and water.
- Plants need a place to grow.
- Thriving plants are plants that are doing very well in their environment.

### Scientific Investigation Skills and Processes
- Ask questions.
- Design and conduct the experiment.
- Create meaning.

### Assessment "Look Fors"
- Students can identify the needs of plants for survival.
- Students can distinguish between surviving and thriving.

### Materials/Resources/Equipment
- Lab coat for teacher
- One lab coat (white adult T-shirt or dress shirt) for each student
- PowerPoint slides of and copies of Handouts 8A (Steps for the Plant Experiment) and 8B (Plant Observation Log)
- Chart paper
- Two plants per group of 3–4 students
- Measuring cup(s)
- Rulers
- Student log books

## Implementing the Lesson

1. Put on lab coats and remind students that they are thinking as scientists and working toward earning their Scientific Investigator Certificate. Today's lesson is the fourth step in the journey toward earning their certificate.

 DOI: 10.4324/9781003238386-15

2. Review the basic needs (food, oxygen, water, and shelter or a place to grow and live) of animals with the students. Ask students, "What happens if these needs aren't met?" (The animals will die.) So, in order to survive, these needs must be met.

3. Tell students that plants, like animals, have certain needs that have to be met in order to survive. Divide the students into pairs and ask them to draw or list the needs of plants to survive (food, oxygen, water, light, and a place to grow and live). Create a class list on a large piece of chart paper. Probe students to elicit these needs for survival. (More advanced students may be able to understand photosynthesis. However, a simple explanation would be that plants use air and light to make their own food.)

4. Now, remind students that the title of the unit is *Survive and Thrive*. Ask students:
   - Now that you know what survive means, what do you think thrive means?

   Discuss with students that when something thrives, whether it is a plant, an animal, or a person, it means that it is doing better than just having its basic needs met. Ask students to think about their environments using the following question:
   - Do you know something, plant or animal, that is doing really well, better than just surviving? Tell about it. (For example, due to conditions, plants may survive in my garden but thrive in Anne's garden.)
   - What would the differences look like? (Use photographs of plants or ecosystems that are surviving and others that are thriving to show the differences.)

5. Begin a class chart of questions the students have about plants surviving versus thriving. What questions would they want answered?

6. During today's lesson, the students will be working with plants to meet these scientific skills: Ask Questions, Design and Conduct the Experiment, and Create Meaning. Use the PowerPoint slide or transparency of Handout 8A as you discuss the setup of today's experiment.

7. Tell students that each group will have two plants that are the same. In thinking about what the plants need to survive, what conditions can students control so that one plant survives and one plant thrives? Explain that we don't want to kill any of the plants. The five variables are oxygen, food (Miracle-Gro® or another fertilizer could be considered, but this would have to be carefully monitored by the teacher), water, light, and a place to live.

8. The students need to decide what variable they want to change in the conditions of the plants (e.g., too much water versus too little). The teacher can assign one variable per group to make certain that all variables are covered. The goal is for students to see the difference between a plant just surviving and one really thriving. (If students are unable to be independent enough to conduct the experiments in small groups, the teacher can guide them through as a whole class.)

9. Talk to students about making a hypothesis, a prediction about what they think will happen. Plants should be labeled so that it is easy for the students to chart the information (see Handout 8B). Students should measure the plants at the beginning of the experiment as a baseline. If students do not know basic measurement skills, the teacher will have to assist the students. A few minutes should be spent every few days measuring, documenting the appearance/changes, and caring for the plants in their specified conditions. After a week or

two, or as soon as changes are significant enough, conclude the experiment. Help the students create meaning from their findings.

10. As students finish designing and setting up their experiments, begin a discussion about change. Review the generalizations of change with the students. Ask the students to carefully consider the different generalizations. They will respond to the following prompt in their log books:
    • Select one generalization of change and tell how it relates to the plant experiment by writing or drawing your answer.

## Concluding and Extending the Lesson

### Concluding Questions and/or Actions
- Share student log book entries.
- What have we learned about change?
- Which generalizations apply?
- What do plants need to thrive?
- Students could take what they have learned about what plants need and attempt to turn one of the "surviving" plants into a "thriving" plant.

### What to Do at Home
- Ask the students to look at the plants around their house or yard and answer this question: "Are the plants surviving or thriving?"
- Ask the students to talk with their families or friends about what they do to care for plants.

Name:_____ Date:_____

# Steps for the Plant Experiment

**Hypothesis:** _____

_____

## Experiment Steps:
1. Each group is given two plants.
2. Label plants **A** and **B**.
3. Each group decides which variable they are going to control: air, water, light, place to grow, or food.
4. Measure the plants and record the measurements.
5. Draw a picture of the plant in the log book.
6. Make a hypothesis about what the group thinks will happen.
7. Water and feed according to the variable for about 10 days or until significant changes are noted. (Water should be measured using the measuring cup and the amount given should be recorded.)
8. Students should record changes in size and appearance in their log books.
9. Measurements should be recorded on Handout 8B.
10. Students should determine whether their hypothesis was correct at the end of the experiment.

## Materials and Resources
- Plants
- Measuring cup
- Water
- Rulers

Name:_____ Date:_____

# Plant Observation Log

| | Plant A | Plant B |
|---|---|---|
| **Day #** _____ | | |
| **Day #** _____ | | |
| **Day #** _____ | | |
| **Day #** _____ | | |

# Lesson 9:
# How Do Animals Look Different?

## Planning the Lesson

### Instructional Purpose
- To identify the different animal body coverings, such as hair, fur, feathers, scales, and shells, and their purpose.

### Instructional Time
- 45 minutes

### Change Concept Generalizations
- Change is related to time.

### Key Science Concepts
- Animals have different body coverings, such as hair, fur, feathers, scales, and shells.
- Body coverings help animals in different ways.

### Scientific Investigation Skills and Processes
- Make observations.
- Learn more.
- Tell others what was found.

### Assessment "Look Fors"
- Students can identify body coverings.
- Students can identify which body coverings go with which animals.
- Students can identify a purpose for body coverings.

### Materials/Resources/Equipment
- Animal flashcards, one set to be divided among the whole class, or, ideally, one set per group of 3–4 students
- Student copies of Handouts 9A (Body Coverings Chart) and 9B (Video Observation Log)
- Chart paper
- *Eyewitness: Pond and River* (video)
- *Marty Stouffer's Wild America* (video)
- *National Geographic: Really Wild Animals—Swinging Safari* (video)
- *National Geographic: Really Wild Animals—Totally Tropical Rainforest* (video)
- Computers with Internet access (if you choose to use webcam sites in addition to video clips)
- Student log books

DOI: 10.4324/9781003238386-16

## Implementing the Lesson

1. Remind students that they are thinking like scientists and are working toward earning their Scientific Investigator Certificate. Today's lesson is the fifth step in the journey toward earning their certificate.

2. Review the Living Things Concept Map (Handout 6G) with students. Ask them to talk about the relationships of living things to patterns, life cycles, and their parents.

3. The *Scholastic Children's Dictionary* (2007) defined a *characteristic* as "a typical quality or feature." Ask students what some of their physical (having to do with the body) characteristics are.

4. Tell students that they will be talking about the physical characteristics of animals (in particular, their body coverings). Students should be divided into pairs. Instruct students to look at their packet of animal cards in their groups and to pay attention to what the animals look like. Allow them a few minutes to look at the cards. Begin a discussion about their observations with the following questions:
   - What is a body covering?
   - What different body coverings do you see?

   Create a class list on a piece of chart paper titled Body Coverings. (Responses should include fur, feathers, hair, scales, skin, and shells.) If students name appendages, start a second chart titled Appendages and use it for Lesson 10. Remind students to focus on Body Coverings today.

5. Give students a copy of Handout 9A. Instruct them to think of different animals that can fit into each category and draw or write different animals for each.

6. Once the information has been compiled on the chart, begin a class discussion by asking the students:
   - Why do we have body coverings?
   - What purposes do they serve?
   - Why do some animals have shells? What are their bodies like?
   - Do the specific body coverings of an animal ever change? (The fable of "The Ugly Duckling" may be referenced or shared here, if there is time.)

7. Tell students that while conducting observations of animals from video clips they will complete the following science skills: Make Observations, Learn More, and Tell Others What was Found. Select clips from this list or other DVDs that you have available in the library:
   - *Eyewitness: Pond and River*
   - *Marty Stouffer's Wild America*
   - *National Geographic: Really Wild Animals—Swinging Safari*
   - *National Geographic: Really Wild Animals—Totally Tropical Rainforest*

   Students will conduct observations of the animals using the video clips. They are to make notes on Handout 9B about what animal(s) they observed and note information about body coverings. The video clips should be watched as a whole class. Then, depending upon their ability, have students complete Handout 9B as a whole class or in pairs. Show several short clips over the remainder of the unit to allow students many observation opportunities.

8. Refer back to the chart paper about body coverings. Ask students:
   - What other changes happen to animals naturally? Probe students until they are able to come up with some examples (e.g., shedding, molting, hibernation).
   - Why do you think a polar bear's fur is white or why does a chameleon change colors? (The animals have camouflage to help them blend in with their environment.)
   - Can you give other examples of camouflage?
   - How does camouflage help animals? (It provides protection from predators.)

9. Have students respond to the following prompt in their log books:
   - Draw and write other examples of animals that use camouflage and explain how it helps them to survive and thrive.

## Concluding and Extending the Lesson

### Concluding Questions and/or Actions
- Share log entries in small groups.
- What is a body covering?
- Why are body coverings important to animals?
- How does camouflage help animals?
- How do animals look different?
- It also is beneficial to provide students with opportunities to observe real animals. If your school setting permits, hang a birdfeeder outside your classroom window or where it can be observed by students. Other possibilities include making an ant farm or having an aquarium in the classroom.
- Look at webcam sites: These sites can be bookmarked for easy access by students and can be used in addition to the video clips. Some sites with animal webcams include:
  o http://www.mbayaq.org/efc/cam_menu.asp
  o http://www.sandiegozoo.org/zoo/ex_panda_station.html
  o http://www.learner.org/jnorth
  o http://www.sandiegozoo.org/zoo/ex_polar_bear_plunge.html
  o http://www.wbu.com/feedercam_home.htm
  o http://nationalzoo.si.edu/Animals/WebCams
  o http://www.museum.vic.gov.au/spidersparlour/tarant.htm
  o http://www.nationalgeographic.com/crittercam/index.html

### What to Do at Home
- Encourage students to talk with their parents about people and camouflage and to ask questions such as, "Do we use it?" "When?" "Why?"

Name:_____ Date:_____

# Body Coverings Chart

List animals in the box underneath the body covering type that the animal has.

| Feathers | Fur | Hair | Scales | Shells |
|----------|-----|------|--------|--------|
|          |     |      |        |        |

Name:_____ Date:_____

# Video Observation Log

| Animal Observed | Observation |
|---|---|
| Animal _____<br><br>Date _____ | |
| Animal _____<br><br>Date _____ | |
| Animal _____<br><br>Date _____ | |
| Animal _____<br><br>Date _____ | |
| Animal _____<br><br>Date _____ | |

# Lesson 10:
# What Is an Appendage?

## Planning the Lesson

### Instructional Purpose
- To understand how animal movement occurs.

### Instructional Time
- 45 minutes

### Change Concept Generalizations
- Change is everywhere.

### Key Science Concepts
- Animals have different appendages, such as arms, legs, wings, fins, and tails.
- Animals move in different ways, such as walking, crawling, flying, climbing, or swimming.

### Scientific Investigation Skills and Processes
- Make observations.

### Assessment "Look Fors"
- Students can match appendages to the correct animals.
- Students can identify which appendages match which movements.

### Materials/Resources/Equipment
- Charts of Handouts 10A (Functions of Appendages) and 10B (Ways Animals Move)
- Copies of Handout 10A, one for each student
- PowerPoint slide, transparency, or copies of Handout 10C (Blank Animal Features Concept Map) for students
- Student log books

## Implementing the Lesson

1. Tell students: "We are going to create a new concept map about some of the features of animals." Show students a PowerPoint slide or transparency of Handout 10C or distribute copies of Handout 10C for student use. Point out the top of the concept map that shows that animals have features. Use the following questions to guide the students in completing the concept map:
    - What is the animal feature that we talked about yesterday? (If the students need a hint, tell them that one of the characteristics of the feature is that it provides protection.)
    - What are some of the body coverings we learned about?
    - How do the body coverings help the animals?

 DOI: 10.4324/9781003238386-17

Use the concept map in Figure 4 as a guide during completion of the concept map.

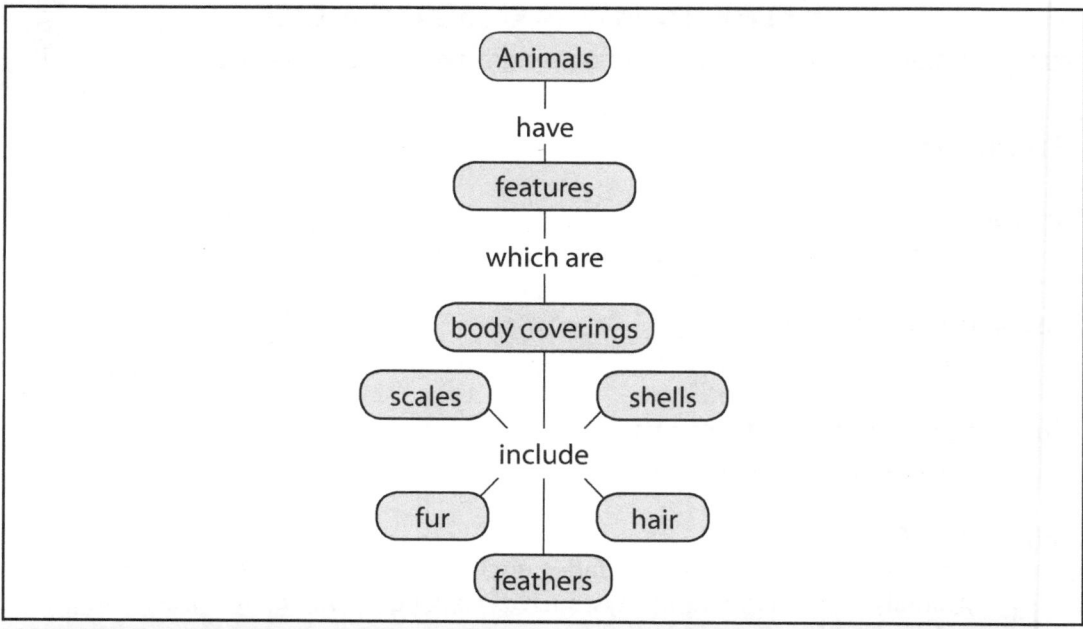

**Figure 4.** Completed concept map.

2. Remind students that they are thinking as scientists and are working toward earning their Scientific Investigator Certificate. Today's lesson is the sixth step in the journey toward earning their certificate. Tell them that today's lesson is about appendages and movement.

3. Post a chart of Handout 10A. Tell students that *appendages* are parts, such as arms, legs, wings, fins, and tails, that extend from the main body and that have specific functions. Engage students in a discussion about the functions of appendages using the following questions. As students respond to the functions of the appendages, record their responses on the chart.
   • What kinds of things can we do with our arms? (We can pick things up and bring them closer or throw them away from us. If students have trouble putting their thoughts into words, have them demonstrate what they mean. They could pick up a pencil or book.)
   • What kinds of things can we do with our legs? (Allow students to demonstrate their meaning by moving or performing the action.)
   • What do animals do with their wings? Fins? Tails?

   Detailed information about the function of different appendages is available in encyclopedias. The information about tails is particularly informative because students may not realize that tails are used in so many different ways by animals.

4. The concept of movement should have come up in the discussion about appendages. If it did not come up, ask students how our arms and legs help us. Post the chart of Handout 10B. Go over the different types of movement with the students. Then, divide the students into pairs. Distribute copies of Handout 10B to the students. Tell the students to list or draw as many animals as they can think of for each type of movement. For example, which animals crawl? After students have had adequate time to come up with answers in their

groups, have the groups take turns sharing their responses and post them on the class chart. Ask students:
- Do some animals move in more than one way (e.g., birds fly but also can walk around on the ground)? Which ones?

5. Continue the discussion about the importance of appendages and the ability to move. Ask students:
   - What are the needs of animals for survival? (food, water, shelter or a place to grow, oxygen)
   - How do appendages help animals survive?

6. Guide students into a discussion about animals without appendages. Ask them:
   - Can you think of an animal that does not have appendages? (Provide hints as needed, such as: they come in many colors such as green, black, and striped; some people are afraid of them; or some are venomous and some are not.)
   - How does a snake move?

Tell them that some snakes, like the rat snake, can climb trees. Have them look at the list of appendages. Ask:
- Does a snake have any appendages? How does it move?
- Can snakes swim? (Most snakes can.)

Tell the students that snakes have been able to adapt to survive in their environments without appendages.

7. Have students respond to the following prompt in their log books:
   - Why are appendages important for animal survival?

## Concluding and Extending the Lesson

### Concluding Questions and/or Actions
- Share student log book entries?
- What are appendages and why are they important for animals?
- What can we learn from animals' movements?
- Encourage students to take home several of the key concept words or pictures for sharing with their families. Ask students to show their families how they can link two of the concepts.

### What to Do at Home
- Have students continue the animal observations of the pet they have been observing.

Handout 10A
# Functions of Appendages

Arms

_____

_____

Legs

_____

_____

Wings

_____

_____

Fins

_____

_____

Tails

_____

_____

Name:_____ Date:_____

# Ways Animals Move

Walking

_____

_____

Crawling

_____

_____

Flying

_____

_____

Swimming

_____

_____

Name:_____ Date:_____

# Blank Animal Features Concept Map

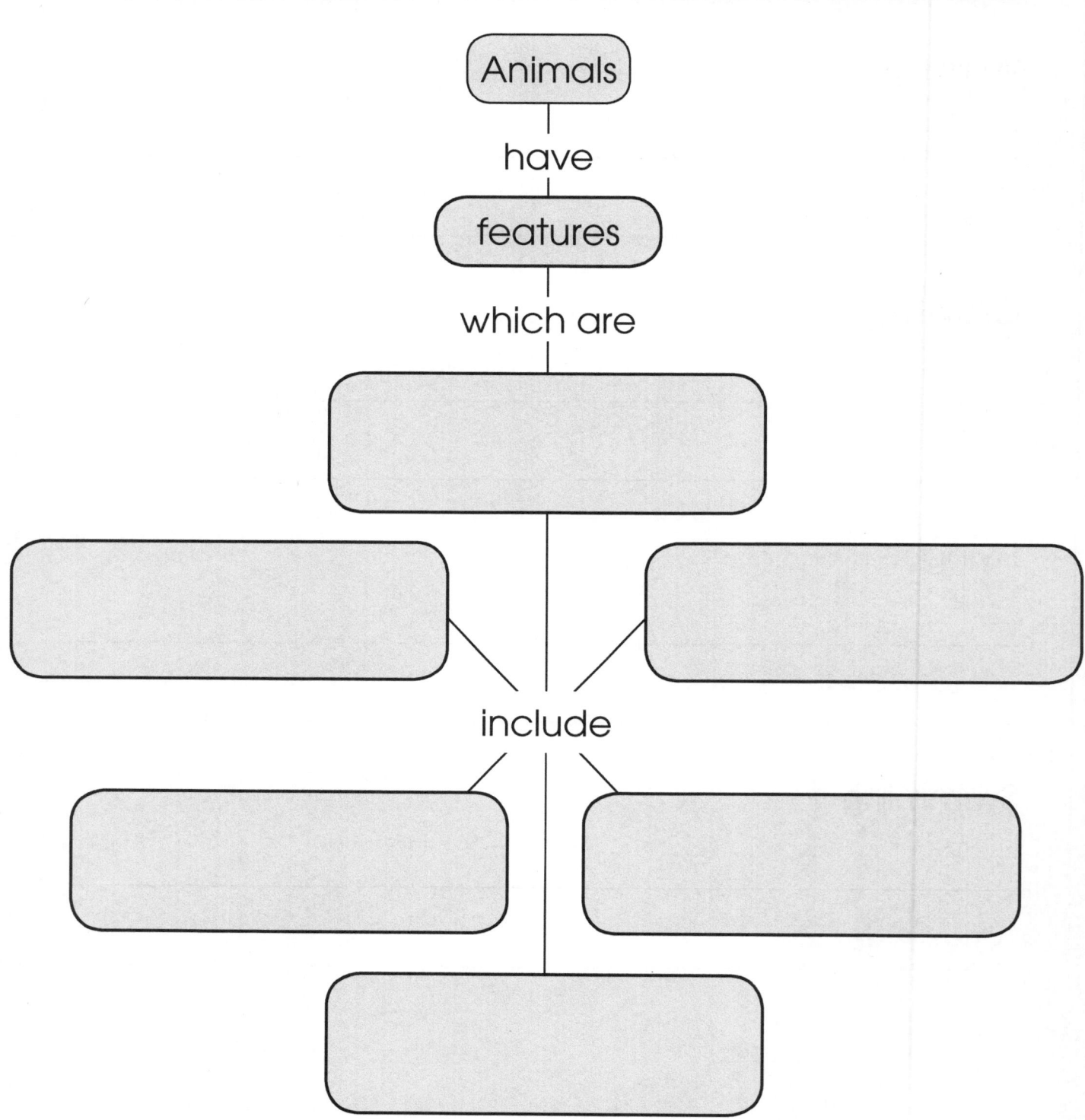

Animals

have

features

which are

include

# Lesson 11:
# How Can We Classify Animals?

## Planning the Lesson

*Instructional Purpose*
- To classify animals in different ways.

*Instructional Time*
- 45 minutes

*Change Concept Generalizations*
- Change is everywhere.

*Key Science Concepts*
- Animals can be classified in different ways.

*Scientific Investigation Skills and Processes*
- Make observations.
- Create meaning.

*Assessment "Look Fors"*
- Students can classify animals in different ways.

*Materials/Resources/Equipment*
- Poster of Handout 11A (Classification Grid for Animals)
- Copies of Handout 11A, one per group of 3–4 students
- PowerPoint slide, transparency, or copies of Handout 11B
- "Tame" and "Wild" written on the board or on chart paper with their definitions
- Pictures of animals from magazines, newspapers, or other sources (if unavailable, students may draw their own pictures during the activity).
- Student log books

## Implementing the Lesson

1. Tell students: "We created a concept map about body coverings, which are some of the features of animals. Today, we are going to focus on some of the other features of animals that we have learned about." Show students a PowerPoint slide or transparency of Handout 11B, or distribute copies of Handout 11B for student use. Point out the top of the concept map that shows that animals have features. Use the following questions to guide the students in completing the concept map:
   - What other features do animals have that help them to survive and thrive?
   - What are some of the types of appendages that animals have?
   - What do appendages allow animals to do?
   - What kinds of movements are typical of animals?

   Use the concept map in Figure 5 as a guide.

 DOI: 10.4324/9781003238386-18

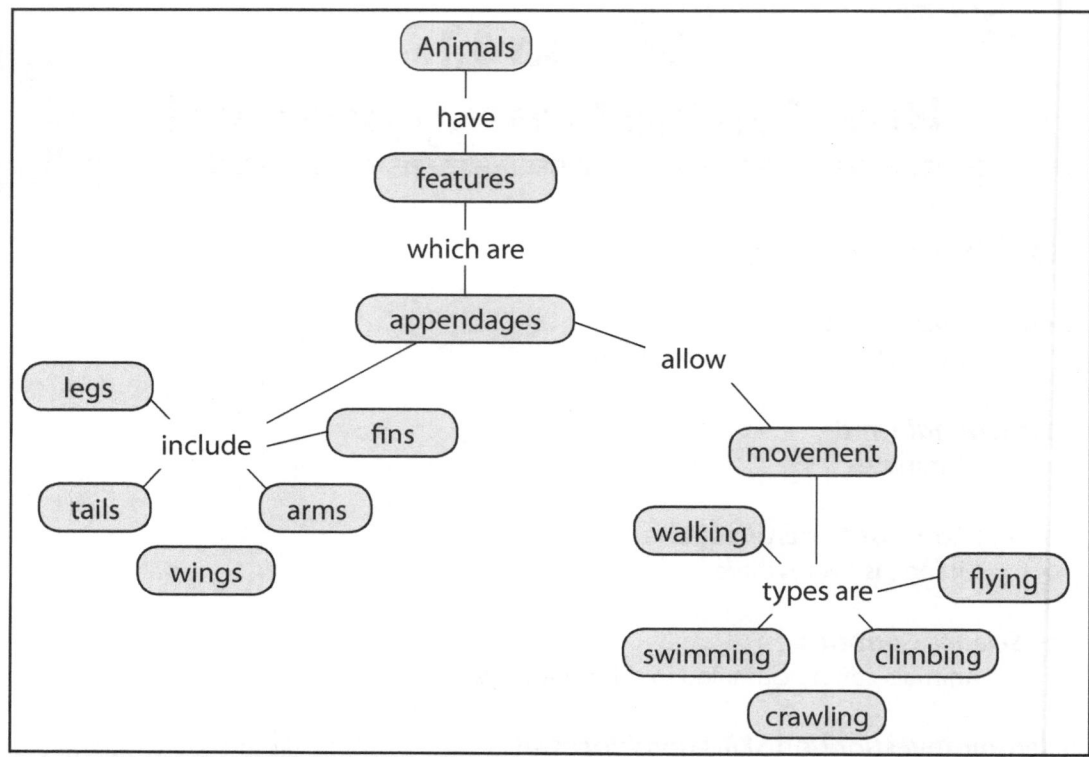

**Figure 5.** Completed concept map for appendages.

2. Remind students that they are thinking as scientists and are working toward earning their Scientific Investigator Certificate. Today's lesson is the seventh step in the journey toward earning their certificate.

3. Students need to be able to classify animals as tame or wild and as living on land or in the water. Ask students:
   - What kinds of pets do we normally have?
   - Why do we choose these types of animals for pets instead of other animals like alligators?
   - So, is an alligator tame or wild?
   - Do we have fish as pets? Where do they have to live?

   Provide additional examples as necessary for students.

4. Follow this discussion up with the following statement: "Animals can be tame or wild and they can live on land or in water." Divide students into pairs and give each group of students a copy of Handout 11A and multiple animal pictures to place in the correct squares on the grid. If multiple pictures are not available, have students draw them on paper. Walk around and monitor the choices the students are making. Discuss their choices with them, asking questions such as:
   - How did you decide where to put _____?
   - Does this animal require shelter? If so, in what kind of shelter does this animal live?
   - Why is this animal considered tame/wild?
   - What might happen if this animal was around people?

5. Have students respond to the following prompt in their log books:
   - Are there other ways to classify animals? How else might we classify animals?

## Concluding and Extending the Lesson

### Concluding Questions and/or Actions
- Share log books.
- How did we classify animals today?
- Why is it important to classify animals?
- How does where an animal lives relate to its characteristics?
- Provide a center activity with a list of linking words and picture cards. Use either general pictures or pictures specific to the unit. Encourage students to practice making mapping sentences by using two cards and connecting them with linking words.

### What to Do at Home
- Ask the students to complete the pet observation log.

Name: _____ Date: _____

# Classification Grid for Animals

|  | Tame | Wild |
|---|---|---|
| Land |  |  |
| Water |  |  |

Handout 11B

# Animal Features Concept Map

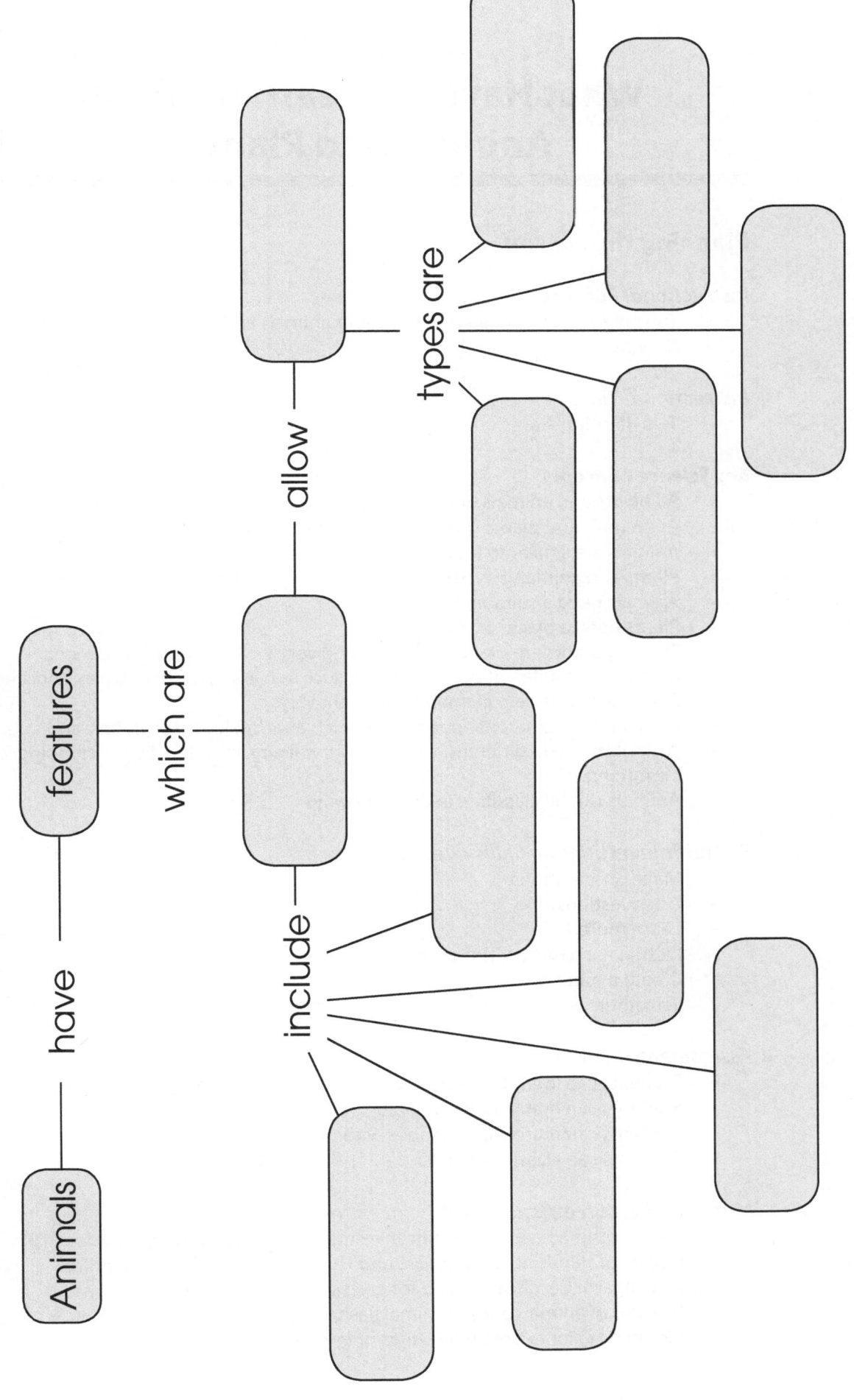

Animals — have — features — which are — [  ] — allow — [  ]

include

types are

# Lesson 12:
# What Have We Learned About Animals and Plants?

## Planning the Lesson

### Instructional Purpose
- To apply what has been learned about animals and plants surviving and thriving.

### Instructional Time
- Two 45-minute sessions

### Key Science Concepts
- All plants and animals undergo different changes in their life cycles.
- As animals and plants grow, they get larger according to a pattern.
- Animals are similar to their parents.
- Plants and animals have basic needs for oxygen, food, and water.
- Animals need a suitable place to live.
- Plants need a place to grow.
- Thriving plants are plants that are doing very well in their environment.
- Animals have different body coverings, such as hair, fur, feathers, scales, and shells.
- Body coverings help animals in different ways.
- Animals have different appendages, such as arms, legs, wings, fins, and tails.
- Animals move in different ways, such as walking, crawling, flying, climbing, or swimming.
- Animals can be classified in different ways.

### Scientific Investigation Skills and Processes
- Make observations.
- Ask questions.
- Learn more.
- Design and conduct the experiment.
- Create meaning.
- Tell others what was found.

### Assessment "Look Fors"
- Students can identify the basic needs of animals and plants.
- Students can match body coverings to groups of animals.
- Students can match appendages with movement.
- Students can classify animals.

### Materials/Resources/Equipment
- Chart of the Wheel of Scientific Investigation and Reasoning (Handout 1B).
- Copies of Handout 12A (Survive and Thrive Cards) and 12B (Scientific Investigator Certificates), one for each student
- Pictures of animals and/or animal flashcards
- Pictures of plants from magazines or Internet printouts

DOI: 10.4324/9781003238386-19     **106**

- Blank paper for each student
- Scissors

## Implementing the Lesson

## Part I

1. Tell students that they are going to review what they have learned about surviving and thriving.
2. Give each student a picture of an animal and a picture of a plant. (The animal flashcards can be used for this activity.) Also give each student a copy of Handout 12A.
3. Instruct students that they are to cut out the cards and match all of the correct cards to their animal and plant, depending on the characteristics.
4. The next step is for the students to look at the cards they have matched to their animal and plant. Students should pick the card or cards that they think helps their animal and/or plant survive the best. (For example, is it the animal's body covering that helps it blend in with the environment, or is it the animal's movements that help it escape predators?) Students should be prepared to share their explanations orally with the class.

### Concluding Questions
- How do animals and plants grow and change during their life cycles?
- What are some of the features of animals that change?
- Select one of the generalizations of change: (1) Change is everywhere. (2) Change is related to time. (3) Change can be natural or manmade. (4) Change may be random or predictable. Tell me something you learned during *Survive and Thrive* that relates to the generalization you selected.

## Part II

1. Point to the Wheel of Scientific Investigation and Reasoning. Ask students to think about the experiments they have done in class: the flower petals, the surviving or thriving plants, and the mealworms. Distribute a piece of paper to each student. (Students can work in small pairs to facilitate this assignment as needed.) Tell them to use pictures and words to describe what they did in the experiment. They can use their log books and the wheel if necessary.
2. Hold a small ceremony to distribute the Scientific Investigator Certificates. Call students up individually and award them their certificates. Encourage each student to share his or her favorite lesson from the unit or what he or she would be interested in studying in science class in the future.

## Concluding and Extending the Lesson

### Concluding Questions and/or Actions
- Why is it important for scientists to study animals?
- How did learning about change help us in our studies of animals?
- Share in small groups: What was your favorite thing you learned about animals?
- Provide each student with a picture that represents some unit concept. Ask the students to work together to put the pictures together in a concept map on chart paper. Help them to make links to build a class concept map, adding words as needed.

# Survive and Thrive Cards

| | | |
|:---:|:---:|:---:|
| Shell | Hair | Fur |
| Feathers | Scales | Wings |
| Tail | Legs | Arms |
| Fins | Swim | Walk |
| Crawl | Climb | Fly |
| Air | Food | Water |
| Shelter | On land | In water |
| Wild | Tame | Has a life cycle |

# Scientific Investigator Certificates

I'm a Scientific Investigator!

Awarded to: _____

Given By: _____

Date: _____

# Postassessment

## Planning the Lesson

### Instructional Purpose
- To assess student knowledge of the concept of change, student understanding of unit content about animals, and student skills in the scientific process.

### Instructional Time
- Macroconcept assessment: 30 minutes
- Content assessment: 30 minutes
- Scientific process assessment: 20 minutes

### Materials/Resources/Equipment
- Copies of Postassessment for Change Concept, Incomplete Animals Concept Map, and Word Bank for Animals Concept Map for each student
- Postassessment for Key Science Concepts, Rubric 1 (Scoring Rubric for Change Concept; see p. 28), Preteaching for Key Science Concepts Postassessment, Sample Concept Map, Rubric 2 (Scoring Rubric for Content Assessment; see p. 33), and Rubric 3 (Scoring Rubric for Scientific Process; see p. 40) for your use
- Copies of Does Sand Dissolve in Water?, What Materials Will You Need?, How Would You Conduct Your Experiment?, What Does This Table Show?, and What Will Dissolve? handouts for the Postassessment for the Scientific Process for each student
- Pencils
- Large chart paper
- Drawing paper for each student

## Implementing the Lesson

1. Each assessment should be given on a separate day.
2. Give each student a copy of the postassessments to complete in the order noted above. The assessments should take no more than 80 minutes.
3. Explain that the assessment will be used to see how much students have learned during the unit.

## Scoring
- Use the rubrics contained in the preassessment sections for concept, scientific process, and content.

# Postassessment for Change Concept

1. What is change? In each box, draw a picture or
   write a word for something that changes.

| | |
|---|---|
| | |
| | |
| | |
| | |

2. Draw a picture of something in your life that changes, and show how it changes. Include as many details as you can.

3. Draw five ways a tree could change or be changed.

# Postassessment for Key Science Concepts

**Directions to Teacher:** Read the following paragraph to the students.

Today I would like you to think about all of the things you know about animals. Think about the connections you can make. You will be completing a concept map, just like the ones you did when we discussed the farm. Look at the word bank and the concept map. You will use some of the word bank words to fill in the parts of the concept map. Some words are just extras that you won't need. Remember, a concept map is used to tell what we know and to make connections. Today's question is: "Tell me everything you know about animals."

## For Kindergarten Students

Direct students to use the word bank to complete the assessment. Students also may use other responses that they come up with on their own. Tell students to draw a picture or write the word or letter for their responses in the appropriate blanks. Each correct response earns one point. Students may enter the word *or* just the letter corresponding to the word *or* come up with their own word.

## For First-Grade Students

Direct students to complete the assessment with appropriate words, pictures, or their own choice of words. Each correct response earns one point.

Name: _____ Date: _____

# Postassessment for Key Science Concepts
## Incomplete Animals Concept Map

Some different animals are:

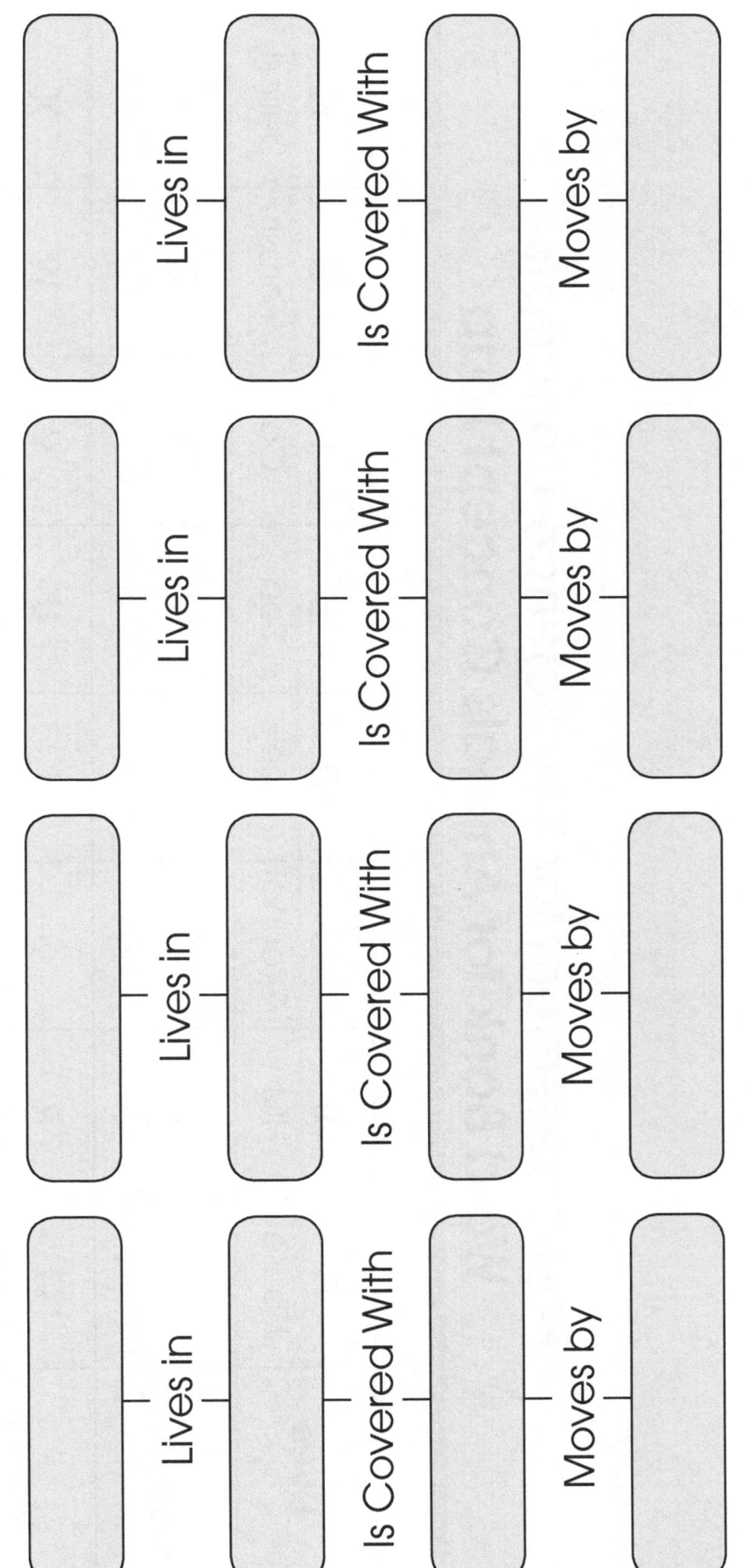

# Postassessment for Key Science Concepts
## Word Bank for Animals Concept Map

| Lives | A<br>House | B<br>Hut | C<br>Ground | D<br>Tree | E<br>Sea | F<br>Car | G<br>Forest | H<br>Desert |
|---|---|---|---|---|---|---|---|---|
| Covering | I<br>Feathers | J<br>Fly | K<br>Hair | L<br>Leaves | M<br>Scales | N<br>Shells | O<br>Fur | P<br>Water |
| Moves | Q<br>Swim | R<br>Walk | S<br>Scales | T<br>Fly | U<br>Tail | V<br>Crawl | W<br>Climb | X<br>Fin |

# Postassessment for the Scientific Process

1. Assess students in groups of 4 to 6.
2. Tell students they are going to think like scientists. Say to students, "I have a scientific question for you: Does sand dissolve in water? You are going to think about whether or not sand dissolves in water. We will work together to look at some pictures and select an answer to some questions about an experiment to find out if sand dissolves in water."
3. Pass out the packet of assessment record sheets on pp. 118–122. Ask students to look at the first sheet (Does Sand Dissolve in Water?). Ask them to write their name on the paper. Direct them to think about the two pictures and make a prediction about whether or not sand dissolves in water. Tell students to put an X in the box under the picture that shows their prediction—sand does not dissolve in water or sand does dissolve in water.
   Picture choices are:
   a. Clear container with water and sand on the bottom
   b. Clear container with water and no sand on the bottom

4. Ask students to think about what materials they will need for their experiment. Look at the What Materials Will You Need? handout (the one that shows some materials that could be used). Ask students to put an X under each picture that shows a material that will be used in the experiment.
   Picture choices are:
   a. Clear container
   b. Spoon
   c. Sand
   d. Salt
   e. Water
   f. Milk

5. Present each student with a set of four cards showing pictures of the steps in the experiment (see the How Would You Conduct the Experiment? handout). Tell the students to select the pictures that show the steps they would take for the experiment. Picture choices are (1) gathering the materials, (2) pouring in water, (3) pouring in sand, (4) stirring the mixture. Instruct students to put the steps they selected in the correct order—which comes first, second, etc. Check to see each student's response and record.
6. Ask students to look at the table on the What Does This Table Show? handout and decide whether it shows that sand dissolves in water or salt does not dissolve in water. Students should put an X in the correct box.
7. Ask students to look at the handout, What Will Dissolve?, with pictures of various materials. The materials are: leaf, twig, salt, JELL-O, crayon, sugar, rock, oatmeal. Direct students to think about things that probably dissolve in water and to place an X in each box under a picture that shows something that will dissolve. Which of these materials will dissolve?

# Does Sand Dissolve in Water?

Does sand dissolve in water? Put an X in the box that matches your prediction.

# What Materials Will You Need?

What materials do you need to conduct your experiment? Put an X in the box of each material you would use.

☐ ☐ ☐

  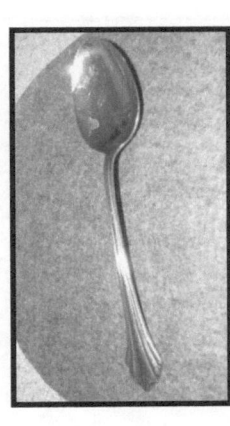

☐ ☐ ☐

Name:_____ Date:_____

# How Would You Conduct
# Your Experiment?

Cut out the pictures below and place them in order of the steps of the water and sand experiment.

# What Does This Table Show?

Did the sand dissolve in water?

| | |
|---|---|
| Cassidy | No |
| Lowell | No |
| Sandy | No |
| Adrian | No |
| Leslie | No |
| Lincoln | No |
| Jonah | No |
| Chwee | No |
| Sun | No |

___ Yes, the sand dissolved in water.

___ No, the sand did not dissolve in water.

Name:_____ Date:_____

# What Will Dissolve?

Put an X in the box below each picture that shows something that will dissolve in water.

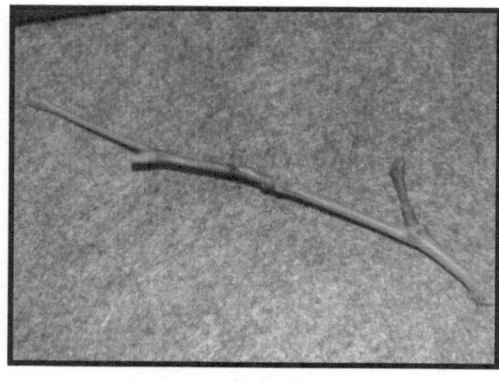 

# Appendix A
# Concept Paper on Change

By Beverly T. Sher, Ph.D.

This paper was adapted from: Sher, B. T. (2004). Change. In J. L. VanTassel-Baska (Ed.), *Science key concepts* (pp. 31–35). Williamsburg, VA: Center for Gifted Education, The College of William and Mary.

We live in a changing world. Change and the absence of change are important features of both scientific and nonscientific processes. Change can occur in simple, predictable ways: Winter gradually gives way to spring, water evaporates from puddles after the rain ends, the sun rises and sets, we (well, most of us, anyway) change from a state of sleep at night to being awake in the daytime. It also can occur in more complicated and unpredictable ways: Hemlines rise and fall apparently randomly, the stock market gyrates, mutations resulting in changed organisms occur, the weather changes. It can also fail to occur: A bottle of root beer in the pantry, when opened, contains the same amount of liquid that it had months earlier at the bottling plant; healthy people maintain roughly the same body temperature at all times (with small, cyclic, daily variations); a rock formation photographed a year ago has the same contours now as it did then. In this concept paper, we will explore the concept of change and the related concept of equilibrium.

There are four general patterns of change. They include:
1. Steady changes: changes that occur at a characteristic rate.
2. Cyclic changes: changes that repeat in cycles.
3. Random changes: changes that occur irregularly, unpredictably, and in a way that is mathematically random.
4. Chaotic changes: changes that appear random and irregular on the surface, but are in fact predictable (in principle).

The first type of change is steady change. Change of this sort occurs at a predictable rate. For example, radioactive decay follows a predictable exponential curve. The number of undecayed atoms remaining after a given time in a radioactive sample can be simply calculated if one knows the half-life of the element involved. For example, half of the atoms in any sample will decay in 14.3 days. Similarly, the distance traveled by a car traveling at 55 miles an hour can be simply calculated for any time after the car leaves (assuming it hasn't run out of gas). Another familiar nonscientific example of steady change would be the steady growth of the balance in an untouched savings account: The rules of compound interest acting on the original balance produce a predictable growth in the daily balance of the account.

The second type of change, cyclic change, frequently is found in nature. A partial list of scientifically interesting cycles would include the phases of the moon, the sunspot cycle, the tides, daily cyclic changes in the levels of hormones in the human body, the sleep-wake cycle in animals, and so on. Some of these examples are familiar to nonscientists as well: We all expect the sun to rise in the east and set in the west once a day, and monthly variations in hormone levels have manifestations that are familiar to almost every female human being over the age of about 13. Other familiar cycles include the credit card billing cycle and the annual recurrence of events such as Christmas, Easter, and fund drives on public television.

The third type of change, random change, also is common in nature. Scientific examples would include the occurrence of spontaneous mutations in genes and the radioactive decay of individual atoms. Nobody can predict when a particular gene in a particular organism will undergo mutation, and nobody can tell by looking at a radioactive atom when it will decay. Interestingly enough, though, these processes that are random at the level of the individual gene or atom have definable rates when one looks at all of the genes in a population of animals (the mutation rate) or all of the radioactive atoms in a sample (the half-life). A similar, nonscientific, example would be winning the lottery. Winning the lottery is a random event. Beforehand, nobody can predict who will win; yet the odds of winning are predictable. Again, this is a process that is unpredictable at the individual level but totally predictable at the level of a population of individuals.

One of the most exciting developments in science in the 1980s was the understanding of chaotic change. Chaos occurs when a system obeys completely predictable behavior (i.e., given the exact state of a system at one time, one can determine its exact state at a future time), but our intrinsic lack of knowledge of its initial state causes its future behavior to appear random. Consider the motion of an asteroid, for example. If you knew the exact position and velocity of an asteroid, you could determine its exact position and velocity at any time in the future. However, a small uncertainty (such as a one millimeter uncertainty in its position) will eventually lead to a huge uncertainty in the future. The position of the asteroid in the future will not be determinable; that is, it will appear random. Similarly, if the location of all of the atoms of our atmosphere were perfectly known at one particular time, the weather could be predicted far into the future. The fact that we don't know those locations exactly (e.g., a butterfly flapping its wings in China will disturb some of the atoms) will lead to huge uncertainties in weather prediction; thus, we will likely never be able to predict the weather very far in advance. It might seem that it is difficult to differentiate between random change and chaotic change, but there are very precise mathematical relations followed by chaotic systems. One of the main realizations of the 1980s was that many systems previously thought to be random, such as water turbulence, weather phenomena, and even cardiac ventricular fibrillation, are in fact chaotic. A good nonscientific example of a chaotic process is the stock market's behavior.

Some systems are characterized not by change but by lack of change. These systems are said to be in equilibrium. The unopened bottle of soda on the shelf in the pantry illustrates this concept in several ways. First, it is static, at rest. All of the outside forces acting on it are balanced: Gravity pulling downward on it is balanced by the force that the shelf exerts to hold it up. Second, the amount of water in it is the same as the amount of water placed in it at the bottling plant. The water in the root beer is not static, however, but is in a state of dynamic equilibrium between two phases: water vapor in the little air space in the neck of the bottle and liquid water in the soda below. The important thing that allows this system to remain in equilibrium is that the bottle is closed. When it is closed, the average number of water molecules that evaporate and leave the liquid phase is exactly balanced by the average number of water molecules condensing from the vapor phase into the liquid phase. Once the bottle is opened, this equilibrium vanishes: Water molecules in the vapor phase can and do escape out the neck of the bottle and the water in the bottle will eventually evaporate. In dynamic equilibrium, therefore, a seeming lack of change reflects balanced processes of change occurring within a system.

The state of dynamic equilibrium that exists in a closed root beer bottle does not depend on any outside forces for its maintenance. Other examples of lack of change, though, represent the interposition of regulatory forces on the system. A simple example of such a system is the thermostat-house system. When the temperature in the house drops below a predetermined level, the sensor in the thermostat notices this and causes the furnace to heat the house. Once the house is warm again, the sensor in the thermostat causes the furnace to shut off. The maintenance of a constant temperature in the house is thus dependent on the regulatory behavior of the sensor. Feedback from the sensor keeps the system stable.

Another, more complicated example of regulated constancy is homeostasis: the maintenance of physical stability within an organism. In humans, for example, everything from our body temperature to the concentrations of different ions in our bloodstream remains fairly constant. This reflects tight control exerted by the cells of our body over these systems: regulation that requires constant sensing of the current state of affairs and compensation for changes that occur as we move, eat, sleep, and do all of the other things that humans do. One example of the mechanisms involved in homeostasis is the action of insulin and its function in regulating blood sugar levels. As a carbohydrate-rich meal begins to be absorbed by the body, the amount of glucose in the bloodstream begins to rise. In response to this, insulin is secreted by the pancreas into the bloodstream. Insulin stimulates the liver to take up extra sugar and store it in the form of glycogen (a starch-like substance); it increases the uptake of sugar by muscle cells and its conversion into glycogen; it inhibits the liver from producing glucose from its glycogen stores; and it stimulates muscle and liver cells to "burn" glucose for energy at a more rapid rate. These activities reduce the amount of glucose in the bloodstream, insulin secretion by the pancreas drops, and things are back to normal. The control of blood sugar levels by the body is complicated and requires a great deal of coordination among the different cells of the body.

## Rationale for Teaching the Concept

Change is an inescapable feature of both scientific and nonscientific systems; indeed, it is often the most interesting feature of either kind of system. In science, for example, the study of developmental biology is concerned entirely with the mechanisms behind the amazing changes that occur as an organism develops from seed or fertilized egg into its mature form; meteorology concerns itself with atmospheric changes; and much of geology involves the study of the changes that have occurred since the Earth was formed. Outside of science, the daily changes in the stock market are important to millions of investors; changes in the weather matter to essentially everyone; and the changes that occur as a baby grows into an adult fascinate parents, grandparents, and teachers alike. An understanding of the basic types of change, as well as of the concept of equilibrium, is useful for anyone.

## Suggested Applications

Many areas of science involve change. Below is a very partial list of suggestions for areas that could be used in illustrating change, equilibrium, and regulation.

### Steady Change

Simple physical changes:
- Temperature change in water as it is heated (and what do you see at the freezing and boiling points?)
- Titration of an acid with a base; watch the pH change
- Rates of random change (mutation formation, radioactive decay)

### Cyclic Change

- Astronomical phenomena (phases of the moon, seasons, changes in day length over the course of the year, behavior of the tides)
- Biological cycles: life cycles, sleep-wake cycle, opening and closing of flowers in plants over the course of the day, the turning of sunflowers to follow the sun, the menstrual cycle

### Random Change

- Radioactive decay
- Spontaneous mutation formation in microbes and man
- Study of probability and statistics

### Chaotic Change

As discussed above, chaotic change is change that could in principle be predicted but, in fact, is unpredictable because of the large numbers of variables involved and uncertainty in measuring starting conditions. Younger children probably have enough difficulty learning that change can be predicted (steady change, cyclic change, rates of random change); introducing chaos at the same time might be confusing. For older children (grades 6–8 or so), the best introduction to chaos would probably be to play with a calculator or computer: Once they have seen the concept in this abstract way, they'll be able to see it in more concrete ways (one of the few cases where hands-on experimentation may *not* be the best way to introduce a new concept!).

### Feedback, Control, and Regulation

- Electronics: study systems like thermostats
- Physiology (human and animal)

# Appendix B
# Teaching Models

## Introduction to the Teaching Models

Several teaching models are incorporated into the Project Clarion units. These models ensure emphasis on unit outcomes and support student understanding of the concepts and processes that are the focus of each unit. Teachers should become familiar with these models and how to use them before teaching the unit. The models are listed below and outlined in the pages that follow.

1. Frayer Model of Vocabulary Development
2. Taba Model of Concept Development
3. Concept Mapping
4. Wheel of Scientific Investigation and Reasoning

## Frayer Model of Vocabulary Development

The Frayer Model (Frayer, Frederick, & Klausmeier, 1969) provides students with a graphic organizer that asks them to think about and describe the meaning of a word or concept. This process enables them to strengthen their understanding of vocabulary words. Through the model, students are required to consider the important characteristics of the word and to provide examples and nonexamples of the concept. This model has similarities to the Taba Model of Concept Development (1962).

In introducing the Frayer Model to your students, demonstrate its use on large chart paper. Begin with a word all students know, such as rock, umbrella, or shoe, placing it on the graphic model. First, ask the students to define the word in their own words. Record a definition that represents their common knowledge. Next, ask students to give specific characteristics of the word/concept or facts they know about it. Record these ideas. Then ask students to offer examples of the idea and then nonexamples to finish the graphic (see Figure B1).

Another way to use the Frayer Model is to provide students with examples and nonexamples and ask them to consider what word or concept is being analyzed. You can provide similar exercises by filling in some portions of the graphic and asking students to complete the remaining sections.

As students share ideas, note the level of understanding of the group and of individual students. As the unit is taught, certain vocabulary words may need this type of expanded thinking to support student readiness for using the vocabulary in the science activities. You may want students to maintain individual notebooks of words so that they can refer back to them in their work.

## Taba Model of Concept Development

Each Project Clarion unit supports the development of a specific macroconcept (change or systems). The concept development model, based upon the work of Hilda Taba (1962), supports student learning of the macroconcept. The model involves both inductive and deductive reasoning processes. Used as an early lesson in the unit, the model focuses on the creation of generalizations about the macroconcept from a student-derived list of created concept examples. The model includes a series

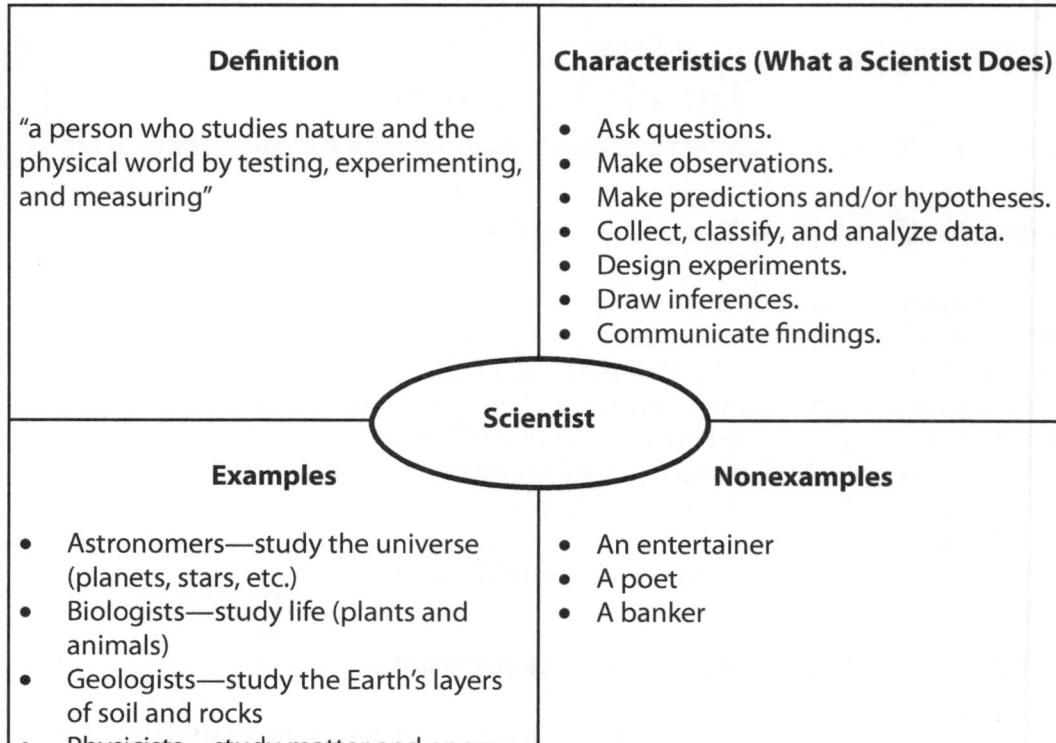

| Definition | Characteristics (What a Scientist Does) |
|---|---|
| "a person who studies nature and the physical world by testing, experimenting, and measuring" | • Ask questions.<br>• Make observations.<br>• Make predictions and/or hypotheses.<br>• Collect, classify, and analyze data.<br>• Design experiments.<br>• Draw inferences.<br>• Communicate findings. |
| **Examples** | **Nonexamples** |
| • Astronomers—study the universe (planets, stars, etc.)<br>• Biologists—study life (plants and animals)<br>• Geologists—study the Earth's layers of soil and rocks<br>• Physicists—study matter and energy | • An entertainer<br>• A poet<br>• A banker |

**Scientist**

**Figure B1.** Completed graphic organizer for Frayer Model.

of steps, in which each step involves student participation. Students begin with a broad concept, determine specific examples of the broad concept, create appropriate categorization systems, cite nonexamples of the concept, establish generalizations based on their understanding, and then apply the generalizations to their readings and other situations.

The model generally is most effective when small groups of students work through each step, with whole-class debriefing following each stage of the process. However, with primary-age students, additional teacher guidance may be necessary, especially for the later stages of the model. The steps of the model are outlined below, using the unit concept of change.

1. Students generate examples of the concept of change, derived from their own understanding and experiences with change in the world. Teachers should encourage students to provide at least 15–20 examples; a class list may be created out of the small-group lists to lengthen the set of changes students have to work with.

2. Students then group their changes into categories. This process allows students to search for interrelatedness and to organize their thinking. It often is helpful to have individual examples written on cards so that the categorization may occur physically as well as mentally or in writing. Students should then explain their reasoning for their categorization system and seek clarification from each other as a whole group. Teachers should ensure that all examples have been accounted for in the categorization system established.

3. Students then generate a list of nonexamples of the concept of change. Teachers may begin this step with the direction, "Now list examples of things that *do not change*." Encourage students to think carefully about their

nonexamples and discuss ideas within their groups. Each group should list five to six nonexamples.

4. The students next determine generalizations about the concept of change, using their lists of examples, categories, and nonexamples. Teachers should then share the unit generalizations and relate valid student generalizations to the unit list. Both lists should be posted in the room throughout the course of the unit.

5. During the unit, students are asked to identify specific examples of the generalizations from their own readings, or to describe how the concept applies to a given situation about which they have read. Students also are asked to apply the generalizations to their own writings and their own lives. Several lessons employ a chart that lists several of the generalizations and asks students to supply examples specifically related to the reading or activity of that lesson.

## Concept Mapping

A concept map is a graphic representation of one's knowledge on a particular topic. Concept maps support learning, teaching, and evaluation (Novak & Gowin, 1984). Students clarify and extend their own thinking about a topic. Teachers find concept mapping useful for envisioning the scope of a lesson or unit. They also use student-developed concept maps as a way of measuring their progress. Meaningful concept maps often begin with a particular question (focus question) about a topic, event, or object.

Concept maps were developed in 1972 by Dr. Joseph Novak at Cornell University as part of his research about young children's understanding of science concepts. Students were interviewed by researchers who recorded their responses. The researchers sought an effective way to identify changes in students' understanding over time. Novak and his research colleagues began to represent the students' conceptual understanding in concept maps because learning takes place through the assimilation of new concepts and propositions into existing conceptual and propositional frameworks.

Concept maps show concepts and relationships between them. (See the sample concept map in Figure B2.) The concepts are contained within boxes or oval shapes and the connections between concepts are represented by lines with linking words.

Concepts are the students' perceived ideas generalized from particular experiences. Sometimes the concepts placed on the map may contain more than one word. Words placed on the line link words or phrases. The propositions contain two or more concepts connected by linking words or phrases to form a meaningful statement.

The youngest students may view and develop concept maps making basic connections. They may begin with two concepts joined by a linking word. These "sentences" (propositions) become the building blocks for concept maps. Older students may begin to make multiple connections immediately as they develop their maps.

As students map their knowledge base, they begin to represent their conceptual understanding in a hierarchical manner. The broadest, most inclusive concepts often are found at the top of a concept map. More specific concepts and examples then follow.

Each Project Clarion unit contains an overview concept map, showing the essential knowledge included in the lessons and the connections students should be able to make as a result of their experiences within the unit. This overview may be useful as a classroom poster that the teacher and students may refer to throughout the unit.

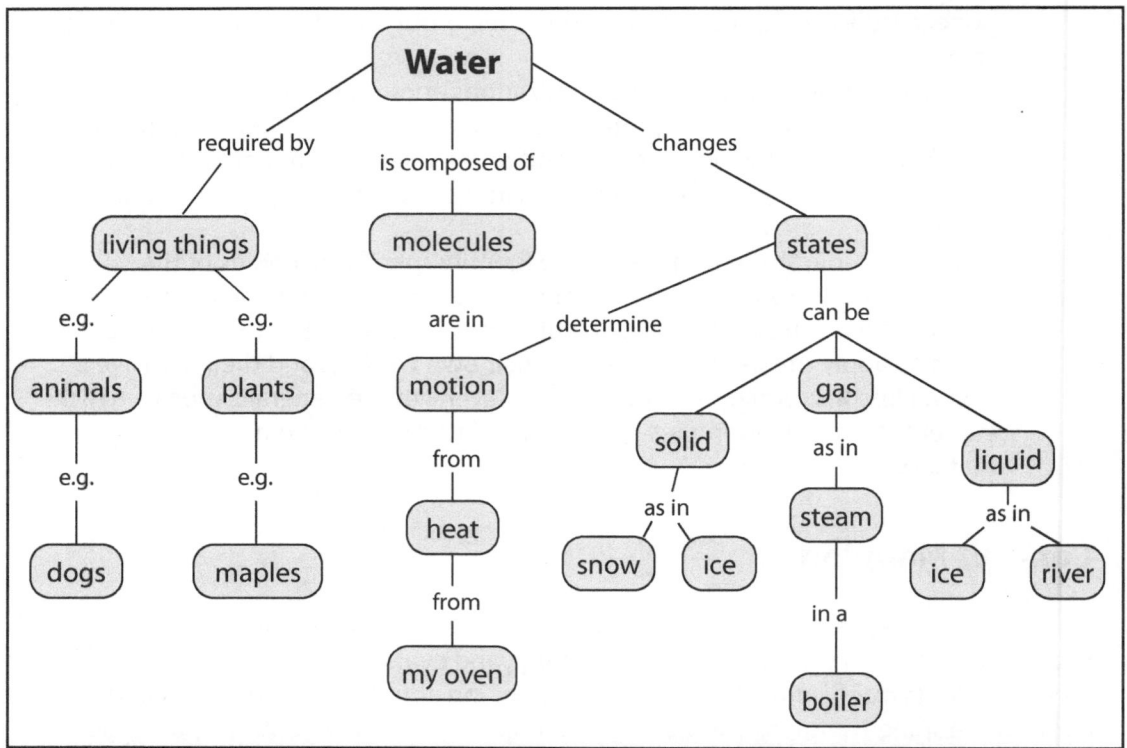

**Figure B2.** A concept map showing a student's understanding of water.

*Note.* Adapted from Novak and Gowin (1984).

## Strategies to Prepare Students for Concept Mapping

The following strategies can be incorporated to help prepare your students for concept mapping activities (Novak & Gowin, 1984).

### What Do Words Mean?

1. Ask students to picture in their minds some common words (e.g., water, tree, door, box, pencil, dog). Start with "object" words, saying them one at a time, allowing time for students to picture each of them.
2. Create a class list of object words, asking students to name other objects they can picture in their minds to add to the list.
3. Next, create a list of event words (e.g., jumping, running, eating). Ask students to envision each of these in their minds and encourage them to contribute to the class list of event words.
4. Give students a few words that are likely to be unfamiliar to most of them, asking if they can see a picture in their mind. These words should be short (e.g., data, cell, prey, inertia). You might include a few simple words in another language. Ask students if they have any mind pictures.
5. Discuss the fact that words are useful to us because they convey meaning. This only happens when people can form a picture in their mind that represents the meaning they connect with the word.

### What Is a Concept?

1. Introduce the word *concept* and explain that concept is the word we use to mean some kind of object or event we can picture in our mind. Refer back to the word lists previously developed as you discuss the word and ask if these

are concepts. Can students see a picture in their mind for each of them? Let students know that when they come upon a word they do not know well enough to form a picture, they will just need to learn the concept associated with that new word.

2. Provide each table with picture cards and ask students to take turns at their table naming some of the concepts included in the card.

### *What Are Linking Words?*

1. Prepare a list of words such as *the, is, are, when, that, then*. Ask students if they can see a picture in their mind for each of these words. Explain that these are not concept words. These are linking words we use when we speak or write to link concept words together into sentences that have special meaning. Ask students if they have any words to add to the list. Label the list "Linking Words."

2. Hold up two picture cards (sky and blue) and give students a sample sentence ("The sky is blue.") Ask students to tell you the concept words and the linking words in your sentence. Give another example.

3. Give each pair of students a few picture cards. Ask the students to work with partners to pick up two cards and then develop a sentence that links the two cards. They should take turns, with one partner making the sentence and the other identifying the concepts and the linking words. Ask them to repeat this a few times and then have several partners share their sentences.

4. Explain to students that it is easy to make up sentences and to read sentences where the printed labels (words) are familiar to them. Explain that reading and writing sentences is like making a link between two things (concepts) they already know. Practice this idea during reading time, asking students to find a sentence and analyze it for concepts and linking words.

## Wheel of Scientific Investigation and Reasoning

All scientists work to improve our knowledge and understanding of the world. In the process of scientific inquiry, scientists connect evidence with logical reasoning. Scientists also apply their imaginations as they devise hypotheses and explanations that make sense of the evidence. Students can strengthen their understanding of particular science topics through investigations that cause them to employ evidence gathering, logical reasoning, and creativity. The Wheel of Scientific Investigation and Reasoning contains the specific processes involved in scientific inquiry to guide students' thinking and actions.

### *Make Observations*

Scientists make careful observations and try things out. They must describe things as accurately as possible so that they can compare their observations from one time to another and so that they can compare their observations with those of other scientists. Scientists use their observations to form questions for investigation.

### *Ask Questions*

Scientific investigations usually are initiated through a problem to be solved or a question asked. Selecting just the right question or clearly defining the problem to be addressed is critical to the investigation process.

### *Learn More*

To clarify their questions, scientists learn more by reviewing bodies of scientific knowledge documented in text and previously conducted investigations. Also, when scientists get conflicting information they make fresh observations and insights that may result in revision of the previously formed question. By learning more, scientists can design and conduct more effective experiments or build upon previously conducted experiments.

### Design and Conduct the Experiment

Scientists use their collection of relevant evidence, their reasoning, and their imagination to develop a hypothesis. Sometimes scientists have more than one possible explanation for the same set of observations and evidence. Often when additional observations and testing are completed, scientists modify current scientific knowledge.

To test out hypotheses, scientists design experiments that will enable them to control conditions so that their results will be reliable. Scientists repeat their experiments, doing it the same way it was done before and expect to get very similar, although not exact, results. It is important to control conditions in order to make comparisons. Scientists sometimes are not sure what will happen because they don't know everything that might be having an effect on the experiment's outcome.

### Create Meaning

Scientists analyze the data that are collected from the experiment to add to the existing body of scientific knowledge. They organize their data using data tables and graphs and then make inferences from the data to draw conclusions about whether their question was answered and the effectiveness of their experiments. Scientists also create meaning by comparing what they found to existing knowledge. The analysis of data often leads to identification of related questions and future experiments.

### Tell Others What Was Found

In the investigation process, scientists often work as a team, sharing findings with each other so that they may benefit from the results. Initially, individual team members complete their own work and draw their own conclusions.

One way to introduce the wheel to students is to provide them with the graphic model (see Figure B3) and ask them to tell one reason why each section of the wheel is important to scientific investigation.

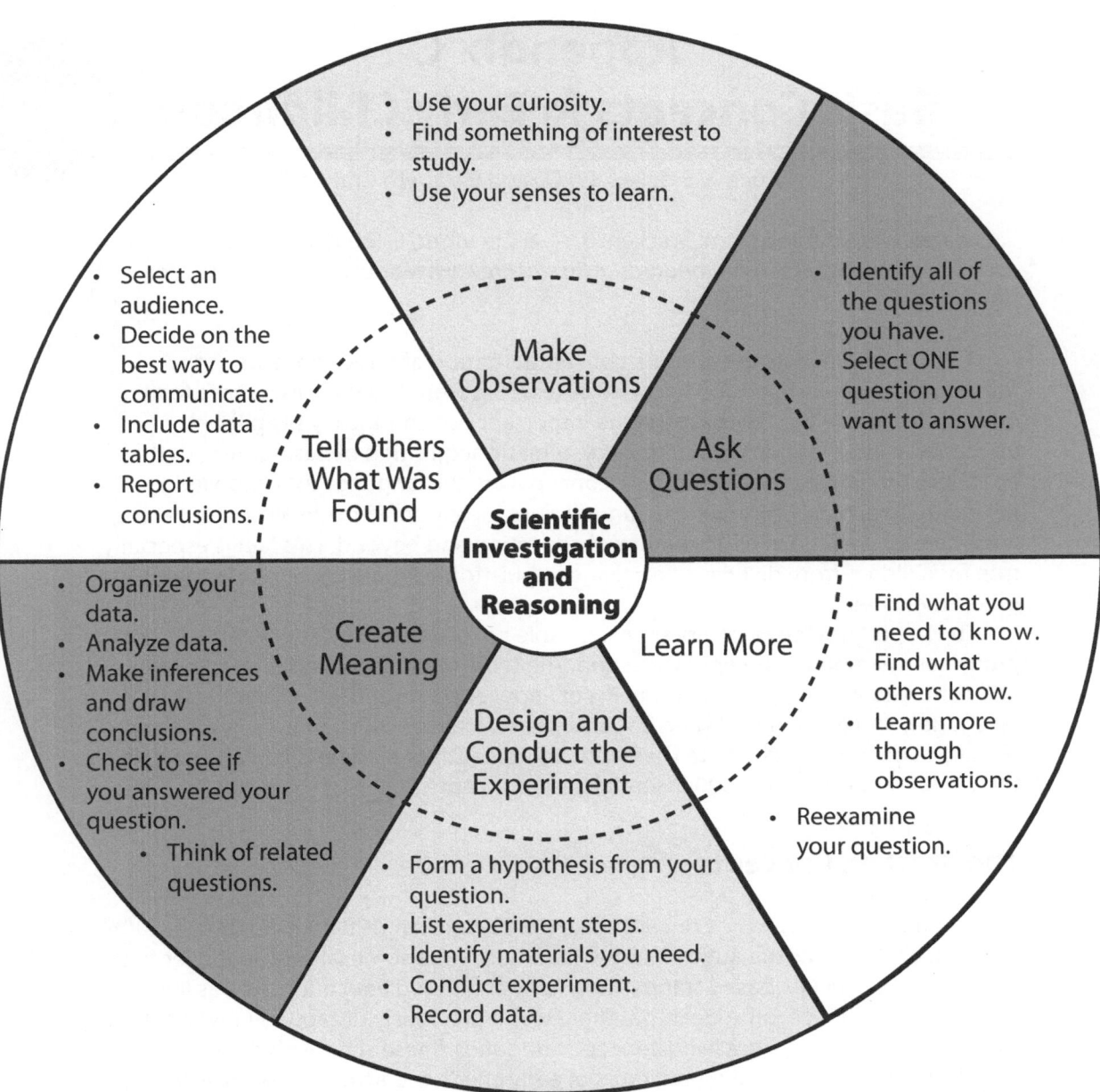

**Figure B3.** Wheel of Scientific Investigation and Reasoning model.

*Note.* Adapted from Kramer (1987).

# Appendix C
# Basic Concepts in Early Childhood

By Bruce A. Bracken, Ph.D., and Elizabeth Crawford

This paper was adapted from: Bracken, B. A., & Crawford, E. (2009). *Basic concepts in early childhood educational standards: A 50 state review*. Manuscript submitted for publication.

This paper presents the authors' conceptualization of basic concepts as the foundation of early childhood knowledge (e.g., Bracken, 1998a). This paper's focus on basic language concepts asserts the importance of empirically supported educational interventions to reinforce systematic acquisition of basic concepts that all children must possess. These basic concepts are the language arts knowledge necessary to explore, comprehend, and discuss topical concepts in all content areas if they are to succeed in early childhood education and beyond. This fact is especially true for children from diverse cultural and linguistic backgrounds, and those with exceptionalities.

The paper provides a comprehensive table of concept knowledge (Table C1) that promotes systematic concept instruction. The table identifies foundational content and conceptual categories, subdomains of knowledge comprised within these categories, and examples of specific concepts, referred to throughout this article as "Bracken concepts" to illustrate the depth and breadth of readiness content revealed in the universe of more than 300 essential basic concepts.

## The Bracken Concepts

When the Bracken Basic Concept Scale (BBCS; Bracken, 1984, 1998b, 2006a, 2006b) was conceived, it was the author's belief that there were some largely unspoken, yet agreed upon, concept-based standards in early childhood education. As this line of work progressed it became clear that there was a previously untapped universal list of essential readiness concepts and concept categories. These school readiness concepts have been shown to be valid predictors of early childhood academic success (Panter, 2000; Panter & Bracken, 2000, in press; Stebbins & McIntosh, 1996; Sterner & McCallum, 1988), cognitive development (Breen, 1985; Howell & Bracken, 1992; McIntosh, Brown, & Ross, 1995; McIntosh, Wayland, Gridley, & Barnes, 1995), language development (Bracken & Cato, 1986; Rhyner & Bracken, 1988); and, they also are ubiquitous within early childhood test directions for early childhood academic and intelligence tests (Bracken, 1986; Bracken & Brown, 2008; Cummings & Nelson, 1980; Flanagan, Alfonso, Kaminer, & Rader, 1995; Kaufman, 1978). Importantly, these concepts can be taught easily, resulting in large educational gains (Wilson, 2004) and their development proceeds along a clear developmental sequence (Bracken, 1988) across both English and Spanish languages (Bracken, et al., 1990; Bracken & Fouad, 1987).

The intent of the Bracken concept list was to identify the universe of basic concepts for parents and teachers so they might more systematically, comprehensively, and effectively teach young children. As such, the Bracken concepts represent one of the first efforts to informally establish early childhood instructional standards. The *Bracken Concept Development Program* (BBCD; Bracken, 1987) was published to provide a direct curricular link between the assessment of children's basic concept knowledge and conceptual instruction. The BCDP presents 19 educationally

# Table C1
# Early Childhood Conceptual Categories

| Concept Category | Subdomain | Concept Examples |
|---|---|---|
| Colors | • Primary Colors<br>• Secondary Colors<br>• Tertiary Colors<br>• Absolutes | • Red, Yellow, Blue<br>• Orange, Green, Purple<br>• Violet, Heather<br>• White, Black |
| Letters | • Recognition<br>  - Uppercase<br>  - Lowercase<br>• Naming<br>  - Uppercase<br>  - Lowercase<br><br>• Letter Sounds<br>• Letter Blend Sounds<br><br>• Letter Production | • Point to M, B, S, D<br>• Point to u, v, c, b<br><br>• Name this letter, W, P, R, E<br>• Name this letter, a, e, g, k<br><br>• What sound does b make?<br>• What sound does ch make?<br><br>• Print the letter X, J, Z |
| Numbers/<br>Counting | • Rote Counting<br>• Place Counting<br><br>• Number Identification<br>  - 0–9<br>  - Double Digits<br>  - Triple Digits<br><br>• Number Naming<br>  - 0–9<br>  - Double Digits<br>  - Triple Digits<br><br>• Number Production<br>• Counting by Sets | • Counting without place value<br>• Counting with one-to-one correspondence<br><br>• Point to the 1, 5, 8, 0<br>• Point to the 22, 58, 95<br>• Point to 138, 395, 783<br><br>• What is this number? 2, 6, 9<br>• What is this number? 44, 78<br>• What is this number? 234, 783<br><br>• Print the number 6, 33, 245<br>• Count to 100 by 2s, 5s, 10s |
| Size/Comparisons | • Three-Dimensional Size<br>• Two-Dimensional Size<br>  - Vertical<br>  - Horizontal<br><br>• Comparative Sizes | • Big, Large, Small, Little<br><br>• Tall, Short<br>• Long, Short<br><br>• Similar, Same, Different |
| Shapes | • Linear (vertical/horizontal)<br>  - Curvilinear Line<br>  - Diagonal Line<br>  - Angular Line<br><br>• Two-Dimensional Shapes<br>• Three-Dimensional Shapes | • Line, Straight<br>• Curve<br>• Diagonal<br>• Angle<br><br>• Circle, Square, Triangle<br>• Sphere, Cube, Pyramid |
| Direction/Position | • Three-Dimensional Direction<br>• Internal/External<br>• Relative Proximity<br>• Self/Other Perspective<br>• Front/Rear<br>• Specific Locations<br>• Cardinal Directions | • Under, Over, Right, Left<br>• Inside, Outside, Around<br>• Near, Far, Beside<br>• My Right, Your Right, My Left, Your Left<br>• In Front of, Behind, Forward, Backward<br>• Edge, Corner<br>• North, South, East, West |

| Concept Category | Subdomain | Concept Examples |
|---|---|---|
| Self-/Social Awareness | • Affective Feeling<br>• Health/Physical<br>• Gender<br>• Familial Relationships<br>• Age<br>• Mores | • Happy, Sad, Excited<br>• Healthy, Sick, Tired<br>• Boy, Girl, Woman, Man, Male, Female<br>• Mother, Father, Brother<br>• Old, Young<br>• Right, Wrong, Correct |
| Texture/Material | • States of Matter<br>• Textures<br>• Materials<br>• Material Characteristics<br>• Temperatures | • Solid, Liquid, Gas<br>• Rough, Smooth, Sharp<br>• Cloth, Wood, Metal<br>• Wet, Dry, Shiny, Dull<br>• Hot, Cold |
| Quantity | • Part/Whole<br>• Relative Quantity<br>• Volume<br>• Multiples<br>• Comparatives/Superlatives<br>• Fractions<br>• Math Signs/Symbols | • Whole, Part, Piece<br>• Lots, Few, Some, None<br>• Full, Empty<br>• Pair, Double, Triple, Dozen<br>• More, Less, Most, Least<br>• Half, One-Third<br>• +, -, x |
| Time/Sequence | • Mathematical Seriation<br>• Frequency<br>• Natural Occurring Events<br>• Temporal Order of Events<br>• Temporal Absolutes<br>• Scheduling<br>• Speed<br>• Relative Age<br>• Temporal Nuances<br>• Larger Temporal Periods | • First, Second, Third<br>• Once, Twice<br>• Morning, Daytime, Evening<br>• Before, After, Finished<br>• Never, Always<br>• Early, Late, Next, Arriving<br>• Fast, Slow<br>• New, Old, Young, Old<br>• Nearly, Just, Waiting<br>• Days, Weeks, Months, Seasons, Years |

sound and empirically supported principles for teaching basic concepts to young children (see Table C2). The Bracken concept list and instructional principles have become important in early childhood assessment and instruction internationally (Bracken, 1984, 1987, 1998b, 2006a, 2006b) because the BCDP integrates this available knowledge and a comprehensive list of basic language concepts into systematic classroom instruction (Bracken, 1987).

A comprehensive discussion of basic concepts is presented below by conceptual categories to bring uniformity to the teaching of basic concepts.

# Colors

Colors are described as primary, secondary, or tertiary, and often are learned by young children in approximately that order. Primary colors are red, yellow, and blue. These colors are considered primary because no combination of colors is blended to produce a primary color. Secondary colors, on the other hand, are colors that result from blending two primary colors. As such, when the two primary colors yellow and blue are combined, they create the secondary color green; when red and blue are combined, they create purple; and when yellow and red are combined, they form orange. Orange, green, and purple then are secondary colors. When primary colors are blended with secondary colors, tertiary or intermediate colors are created, which vary depending on the proportions of each color added to the admixture (e.g., blue and green combined form the tertiary colors blue-green, heather, aquamarine, teal, and so on depending on the proportions of blue or green added).

# Table C2
# Instructional Principles for Teaching
# the Bracken Basic Concepts

| | Bracken Instructional Principle |
|---|---|
| 1 | Language, examples, materials, and procedures used to teach concepts should be less complex than the concept being taught. |
| 2 | When concepts occur in pairs (e.g., up, down) or in series (e.g., before, just, after), maximize the meaningfulness of each concept by teaching all relevant concepts during the same lesson. |
| 3 | As much as possible, teach simple concepts, conceptual pairs, and series by using mnemonic strategies that facilitate understanding and enhance memory. |
| 4 | Concept generalizations should be taught initially by instruction with obvious examples of the concept and proceed to less obvious, more extreme examples. This instructional format should be followed with cases in which "nonexamples" are used to teach concept discrimination. Nonexamples should range initially from the apparent to relative nuances in later lessons. |
| 5 | Identify the characteristics that define the concept, distinguish which single dimension or group of characteristics are most salient, and provide instruction that initially emphasizes the most important characteristics, while minimizing the less important or irrelevant dimensions. |
| 6 | Instruction of polar concepts or concepts in a continuum should begin with the positive pole concept. |
| 7 | Concept pairs should be taught so that children identify positive examples as being *the concept* and negative examples (nonexamples) as *not being the concept*. Objects, for example, are either *tall* or *not tall*. |
| 8 | Once the positive pole concept is accurately described as *the concept* or *not the concept*, the child is taught that when it is *not the concept* it is the *negative pole concept* (e.g., if the object is *not tall*, then it is *short*). |
| 9 | When both polar concepts are learned, the teacher continues to display the logic that if it is *not the positive concept*, then it is *the negative concept* and if it is *not the negative concept*, then it is *the positive concept* (e.g., if the object is *not short*, then it is *tall*). |
| 10 | Consider the sequence in which concepts are acquired; the teacher should continually teach and assess to ensure that concept instruction is at the appropriate level. |
| 11 | School instruction should provide parents with a list of concepts and helpful suggestions as to how concepts can best be taught at home. |
| 12 | Conceptual lessons should elicit active participation and allow for multisensory instructional presentations. |
| 13 | Allow for overlearning in concept instruction by incorporating previously learned concepts in the lessons designed to teach new concepts. |
| 14 | Keep concept instructional sessions appropriately brief. |
| 15 | To ensure overlearning of concepts, allow for an adequate review of previously learned concepts before proceeding to new concepts. |
| 16 | Begin instructional sessions at a level that ensures success. Maintain an instructional difficulty level that guarantees continued success. |
| 17 | Structure conceptual instruction sessions so that each has an identifiable beginning and ending and objectives are clear. |
| 18 | Concepts should be taught in familiar situations in order to facilitate generalization. |
| 19 | To ensure a thorough understanding of basic concepts as instruction progresses, sessions should include conceptual combinations that are more complex than the instruction of single concepts. |

*Note.* Adapted from Bracken (1987).

In addition to primary, secondary, and tertiary colors, there are the additional absolute colors of white and black. From a natural beam of light perspective, white is the combination of all primary colors, colors that can be separated into the full color spectrum comprised in a prism array. From an artificial, projected light beam perspective, white is the combination of red, green, and blue. Also from a light beam perspective, black is the total absence of color or as an extension, the absence of light. From a materials perspective, however, white is the absence of any color pigmentation and black is the combination of all colors. As such, white and black are contributors to the lightening or darkening of primary or secondary colors by degree of addition to the color admixture.

In combination, primary and secondary colors with the absolute colors of white and black added are universal colors for all people with normal color vision and should constitute the educational basis for standards in color recognition and naming (Bracken, 1984, 1998b, 2006a, 2006b). Young children should be able to describe objects in terms of color, including the most basic primary and secondary colors, plus white and black.

## Letters

Recognizing and naming the 26 letters of the alphabet appears to be the very foundation upon which reading preliteracy skills are developed. Developmental literature and the difficulty levels achieved among the Bracken concepts concur that children reliably recognize uppercase (i.e., capital) letters before they recognize lowercase letters, and later they are also able to name uppercase before lowercase letters, and later still they are able to reproduce the sounds that individual letters and consonant blends make.

The Bracken concept list includes the prereading concepts to include important phonemic awareness skills and abilities (i.e., letter and initial consonant blend sounds). Ideally, standards, curriculum, and instruction would systematically follow the developmental sequence of recognition followed by expression, including: (1) identifying uppercase letters, (2) identifying lowercase letters, (3) uppercase letter naming, (4) lowercase letter naming, (5) letter-sound production, and (6) initial consonant blend production.

## Numbers/Counting

As with prereading skills, premath and early math skills have a fairly predictable developmental progression. Early on, young children develop a sense of quantity (e.g., more/less) and develop the ability to rote count without a one-to-one number/object correspondence. Later, young children learn to recognize numbers 1–5, followed by 6–9 and zero, and then double-digit numbers. Along the way, young children begin to count to 10 with one-to-one correspondence, and quickly they are on to counting to numbers greater than 100. Later still, young students learn to count by twos, fives, 10s, and so on.

## Sizes/Comparisons

Sizes and comparative knowledge about size can be thought of in a number of ways, including considering objects in terms of their overall, three-dimensional size (e.g., big, small, large, little) or two-dimensional size, which may be depicted as vertical (e.g., tall, short) or horizontal (e.g., short, long), or diagonal. The developmental literature and item difficulty levels on the BBCS-R3 (Bracken, 2006a) generally support the assertion that students first learn concepts related to gross, three-dimensional size (e.g., big, small) before learning concepts related to two-dimensional size (e.g., tall, long). Students must be able to discern similarities and differences between the many attributes or dimensions of objects in our environment, including dimensions of relative size (e.g., same, equal, different),the more basic Bracken concepts of equal and unequal, and concepts in their most basic form (e.g., short), as well as in their comparative and superlative forms (i.e., shorter, shortest). Additional concepts that provide a more complete list of size concepts include unique contexts (e.g., deep/shallow, thin, thick) or employ comparative size language (e.g., same, not the same, equal, unequal, match, exact, similar).

## Shapes

At the most basic level, shapes begin with lines, which may be straight, curvilinear (i.e., curved), or angled. Lines also may run in vertical, horizontal, or diagonal orientations.

Lines may be connected to create a whole object with two-dimensions (e.g., circle, square) or three-dimensions (e.g., sphere, cube). The comprehensive Bracken list includes many concepts such as those that define line nature (e.g., straight, curve, diagonal, angle), as well as a full range of two- and three-dimensional shapes (e.g., diamond, curve, angle, heart, checkmark, column, row, diagonal).

## Direction/Position

Directions and locations (or positions) are relational concepts that describe the relative location or position of objects in space. From an early developmental orientation (i.e., non-perspective-taking orientation), objects are viewed in their locations from the perspective of the child (e.g., right is from the child's right-hand perspective); older children with the ability to take another's perspective can view locations from the orientation of others (e.g., opposing orientation where Sally's right is understood as the student's left). From a basic knowledge point of view, directional concepts are first learned from a self-perspective orientation and then later from another's perspective.

In addition to perspective, directions and position concepts by and large are represented most frequently as prepositions, but they also may include nouns (e.g., edge, corner). Early directional knowledge emphasizes a three-dimensional orientation from a self-perspective, and includes concepts that address vertical (e.g., above, below, up, down, under, over, high, low, top, bottom), horizontal (e.g., right, left, beside, next to, sideways), three-dimensional (e.g., around, through), internal/external (e.g., in, out, inside, outside, between), relative proximity (e.g., near, close, far), and the child's front or rear (e.g., front, back, forward, backward).

The Bracken concept list includes all of the aforementioned concepts plus many other related directional or positional concepts (e.g., falling, rising, together, apart, side, toward, away, apart, joined, together, height, length, opposite, level, space, moving, still, beginning, end, open, closed, on, off, upside down, following, ahead, behind). The Bracken Concept Development Program provides a comprehensive, logical extension of knowledge and a systematic treatment of the given universe of content.

## Self- and Social Awareness

The domain of self- and social awareness includes a wide array of personological and sociological knowledge, including affective feelings, health and physical condition, gender awareness, familial relationships, relative age, and social mores or correctness. As with academic content areas, students' sense of self and developing self-concepts are developmental in nature (Bracken, 1996).

The Bracken concept list includes the most comprehensive collection of concepts and knowledge in the area of self- and social-awareness. The Bracken concepts include conceptual knowledge associated with gender (e.g., male, female, boy, girl), familial relations (e.g., brother, sister, mother, father), age (e.g., old, young), health and physical awareness (e.g., tired, fatigued, rested, healthy, hurt, relaxing, sleepy, sick), affective state (e.g., happy, sad, crying, laughing, smiling, angry, afraid, excited, frowning, worried, curious), and social mores (e.g., right, wrong, correct, easy, difficult).

## Texture/Material

From a developmental perspective, young children from birth begin to learn directly about their environments, including the attributes that define or characterize the objects in their environments. As infants crawl and toddlers toddle about and handle objects, they begin to develop an awareness of different textures (e.g., rough, hard, soft, smooth) and material characteristics or conditions (e.g., heavy, wet, dry, light). Parents begin to teach their children at very early ages the safety concept of hot and by comparison the polar opposite concept of cold. Much later, children begin to learn what the objects in their environments are made of (e.g., wood, metal, glass, cloth) and they relate to the textures and material attributes that are consistent with each material (e.g., wood is hard; metal is heavy; glass is clear or sharp; cloth is soft, or sometimes rough). Finally, children learn about the manmade changing states of objects or materials (e.g., rough wood can be sanded smooth) or natural changing states of objects and materials (e.g., water can be found in various states, depending on temperature [i.e., liquid, solid, gas]). Such a comprehensive consideration and treatment of materials and textures as conceptual knowledge ensures that children are better able to use their five senses to identify, name, and discriminate between various object attributes, characteristics, and qualities at a young age.

The Bracken concept list includes conceptual knowledge across each of the five senses, except taste. Within the remaining four senses, however, the Bracken concepts comprehensively include knowledge of materials (e.g., cloth, wood), material attributes (e.g., wet, dry), material textures (e.g., rough, smooth, sharp), states of matter (e.g., liquid, solid, gas), temperature (e.g., hot, cold, boiling), sound (e.g., loud, quiet), and appearance (e.g., shiny, bright, clear, dull, dark, light).

## Quantity

Quantitative knowledge in early childhood is part of, yet distinct from, students' understanding of numbers and counting. Knowledge of numbers and counting provides the foundation for much of the quantitative understanding that follows, but not always so. For example, it is obvious that virtually all young children have acquired the concept of more before they can identify numbers or count.

Quantitative concepts, then, represent the understanding of such conditions as part/whole (e.g., whole, part, piece), relative quantity (e.g., lots, few, many, nothing, none, every), volume (e.g., full, empty), comparatives (e.g., more than, less than), multiples (e.g., double, pair, couple, triple, dozen), fractions (e.g., half, third), currency (e.g., dime, nickel, quarter), and the use and understanding of mathematical signs (e.g., +, -, =). Quantitative concepts provide young children with language that allows them to talk about numbers and counting in ways that communicate and generalize knowledge beyond the number of the objects being measured, weighed, counted, divided, distributed, or otherwise treated mathematically.

## Time/Sequence

Because life progresses temporally, from birth to death, from morning to night, from breakfast to dinner, from new to old, from yesterday to tomorrow, young students quickly attend to the temporal patterns in their lives, even if they have not acquired the language to describe those patterns. In the domain of time/sequence, there is the obvious mathematical/quantitative nature of seriation (e.g., first, second, third) and frequency (e.g., once, twice) that also must be considered.

Knowledge of time and sequence, however, is more than just a quantitative component. Time and sequence also deal with students' knowledge and awareness of natural events (e.g., morning, daytime, night), temporal order of events (e.g., starting, before, after, over, finished), temporal absolutes (e.g., never, always), scheduling (e.g., early, late, next, arriving, leaving), speed (e.g., fast, slow), relative age (e.g., new, old, young, old), and descriptive temporal nuances (e.g., nearly, just, quit, waiting).

All of the previously mentioned time/sequence related concepts are found on the Bracken concept list.

## Conclusions

The collective developmental and educational literature and the efforts of individual researchers have identified a comprehensive and unified combination of foundational knowledge that young children should know in order to ensure that all children possess a common knowledge base before entering advanced grades. This foundational knowledge is necessary to ensure that students have the language and understanding to learn about, talk about, and ask about content they learn in social studies, science, language arts, art, mathematics, and so on. This complete list of content and concepts constitute an extremely important foundation of knowledge. If all young children possessed a thorough understanding of the basic concepts subsumed by these overarching categories of content that describe and comprise this universe of basic knowledge, all students would start their formal educations on a much more even footing. The knowledge base included in the list of Bracken concepts and the instructional principles provides parents and teachers a real, nonrhetorical, practical, and proven guide for placing a solid, common, and important foundation under all young students.

# Appendix D
# Materials List

| Lesson | Materials Needed |
|---|---|
| Lesson 1: What Is a Scientist? | • Lab coat for teacher<br>• One lab coat (white adult T-shirt or dress shirt) for each student<br>• Beaker<br>• Microscope or magnifying glass<br>• Prepared charts for students, PowerPoint slides, or transparencies of Handouts 1A (Defining Scientists) and 1B (What Scientists Do: The Wheel of Scientific Investigation and Reasoning)<br>• Poster of the Wheel of Scientific Investigation and Reasoning<br>• Marker<br>• One piece of chart paper<br>• Student log books<br>• *What Is a Scientist?* by Barbara Lehn |
| Lesson 2: What Is Change? | • Chart of Handout 2A (Five Senses Chart)<br>• Copies of Handout 2B (Change Is Everywhere), one for each student<br>• Three or four oranges<br>• One paper cup<br>• Drawing paper<br>• Four sentence strips with a different change generalization written on each strip<br>• Chart paper (to make Taba concept model chart)<br>• Student log books<br>• *My Five Senses* by Aliki (optional: use if background is needed about the five senses) |
| Lesson 3: What Scientists Do—Observe, Question, Learn More | • Lab coat for teacher<br>• One lab coat (white adult T-shirt or dress shirt) for each student<br>• Chart, PowerPoint slide, or transparency of Handout 1B<br>• Chart, PowerPoint slide, or transparency of Handout 3A (Observations of Two Flowers)<br>• Two types of flowers for each pair of students<br>• Log book page for each student using Handout 3A or drawing paper for each student<br>• Chart paper<br>• Student log books |
| Lesson 4: What Scientists Do—Experiment, Create Meaning, Tell Others | • Lab coat for teacher<br>• One lab coat (white adult T-shirt or dress shirt) for each student<br>• Chart of Handout 1B<br>• Charts, PowerPoint slides, or transparencies of Handouts 4A (Definition of Hypothesis), 4B (Steps for Flower Experiment), and 4C (Flower Experiment Data Table)<br>• Copies of Handout 4C, one for each student<br>• One badge per student using Handout 4D<br>• Two types of flowers with a few clearly defined petals (one flower of each type per pair of students)<br>• Chart paper<br>• Pencils<br>• Student log books |

| Lesson | Materials Needed |
|--------|------------------|
| Lesson 5: What Is a Life Cycle? | • One copy of Handout 12B (Scientific Investigator Certificate)<br>• Charts, PowerPoint slides, or transparencies of Handouts 5A (Life Cycle of a Plant) and 5B (Life Cycle of a Frog)<br>• Copies of Handouts 5B and 5C (Personal Life Cycle), one for each student<br>• Scissors<br>• Glue<br>• Chart paper<br>• Markers<br>• Construction paper |
| Lesson 6: What Is the Life Cycle of a Mealworm? | • Lab coat for teacher<br>• One lab coat (white adult T-shirt or dress shirt) for each student<br>• Chart of Handout 1B<br>• Charts, PowerPoint slides, transparencies, or posters of Handouts 6A (Mealworm Investigation Hypothesis) and 6B (Steps for Mealworm Experiment)<br>• PowerPoint slide, transparency, or poster of Handout 6E (Completed Life Cycle of a Mealworm, to be used as a review following the experiment but before the assessment)<br>• Charts, PowerPoint slides, posters, or copies of Handouts 6C (Mealworm Data Table), 6D (Experimental Report Form), 6F (Life Cycle of a Mealworm), 6G (Living Things Concept Map), and 6H (Animals Observation Sheet), one for each student<br>• Live mealworms (may be purchased online, at a pet store, or bait stores often in containers of 50)<br>• Clear plastic shoe box with loose-fitting lid<br>• Oatmeal, bran meal, or wheat meal<br>• Apples or potatoes<br>• Magnifying glasses<br>• Rulers<br>• *Mealworms* by Donna Schaffer<br>• Student log books |
| Lesson 7: What Are the Requirements for Life? | • Lab coat for teacher<br>• One lab coat (white adult T-shirt or dress shirt) for each student<br>• PowerPoint slide or transparency of Handout 6E (Complete Life Cycle of a Mealworm) or another picture of a worm<br>• PowerPoint slide or transparency of Handout 5B (Life Cycle of a Frog) or another picture of a frog<br>• PowerPoint slides of and copies of Handout 7A (Blank Requirements for Life Concept Map)<br>• Several pieces of chart paper; one piece divided into a T with the headings: Needs/Wants<br>• Cup of water<br>• Food items (basic needs and wants)<br>• A few toys<br>• Variety of animal books<br>• Animal Flash Cards (http://www.trendenterprises.com) or a variety of animal pictures<br>• Student log books |
| Lesson 8: What Do Plants Need to Thrive? | • Lab coat for teacher<br>• One lab coat (white adult T-shirt or dress shirt) for each student<br>• PowerPoint slides of and copies of Handouts 8A (Steps for the Plant Experiment) and 8B (Plant Observation Log)<br>• Chart paper<br>• Two plants per group of 3–4 students<br>• Measuring cup(s)<br>• Rulers<br>• Student log books |

| Lesson | Materials Needed |
|---|---|
| Lesson 9: How Do Animals Look Different? | • Animal flash cards, one set to be divided among the whole class, or, ideally, one set per group of 3–4 students<br>• Student copies of Handouts 9A (Body Coverings Chart) and 9B (Video Observation Log)<br>• Chart paper<br>• *Eyewitness: Pond and River* (video)<br>• *Marty Stouffer's Wild America* (video)<br>• *National Geographic: Really Wild Animals—Swinging Safari* (video)<br>• *National Geographic: Really Wild Animals—Totally Tropical Rainforest* (video)<br>• Computers with Internet access (if you choose to use webcam sites in addition to video clips)<br>• Student log books |
| Lesson 10: What Is an Appendage? | • Charts of Handouts 10A (Functions of Appendages) and 10B (Ways Animals Move)<br>• Copies of Handout 10A, one for each student<br>• PowerPoint slide, transparency, or copies of Handout 10C (Blank Animal Features Concept Map) for students<br>• Student log books |
| Lesson 11: How Can We Classify Animals? | • Poster of Handout 11A (Classification Grid for Animals)<br>• Copies of Handout 11A, one per group of 3–4 students<br>• PowerPoint slide, transparency, or copies of Handout 11B<br>• "Tame" and "Wild" written on the board or on chart paper with their definitions<br>• Pictures of animals from magazines, newspapers, or other sources (if unavailable, students may draw their own pictures during the activity).<br>• Student log books |
| Lesson 12: What Have We Learned About Animals and Plants? | • Chart of the Wheel of Scientific Investigation and Reasoning (Handout 1B).<br>• Copies of Handout 12A (Survive and Thrive Cards) and 12B (Scientific Investigator Certificates), one for each student<br>• Pictures of animals and/or animal flashcards<br>• Pictures of plants from magazines or Internet printouts<br>• Blank paper for each student<br>• Scissors |

# References

Bracken, B. A. (1984). *Bracken Basic Concept Scale.* San Antonio, TX: Harcourt Assessments.

Bracken, B. A. (1986). Incidence of basic concepts in the directions of five commonly used American tests of intelligence. *School Psychology International, 7,* 1–10.

Bracken, B. A. (1987). *Bracken Concept Development Program.* San Antonio, TX: Harcourt Assessments.

Bracken, B. A. (1988). Rate and sequence of positive and negative pole concept acquisition. *Language, Speech, and Hearing Services in the Schools, 19,* 410–417.

Bracken, B. A. (1996). Clinical applications of a multidimensional, context-dependent model of self-concept. In B. A. Bracken (Ed.), *Handbook of self concept: Developmental, social, and clinical considerations* (pp. 463–505). New York, NY: John Wiley and Sons.

Bracken, B. A. (1998a). Basic concept acquisition and assessment: A celebration of our world's many dimensions. *Clinicians' Forum, 8*(2), 1, 7.

Bracken, B. A. (1998b). *Bracken Basic Concept Scale—Revised.* San Antonio, TX: Harcourt Assessments.

Bracken, B. A. (2006a). *Bracken Basic Concept Scale—Receptive Third Edition.* San Antonio, TX: Harcourt Assessments.

Bracken, B. A. (2006b). *Bracken Expressive.* San Antonio, TX: Harcourt Assessments.

Bracken, B. A., Barona, A., Bauermeister, J. J., Howell, K. K., Poggioli, L., & Puente, A. (1990). Multinational validation of the Bracken Basic Concept Scale. *Journal of School Psychology, 28,* 325–341.

Bracken, B. A., & Brown, E. F. (2008). Early identification of high-ability students: Clinical assessment of behavior. *Journal for the Education of the Gifted. 31,* 403–426.

Bracken, B. A., & Cato, L. A. (1986). Rate of conceptual development among deaf preschool and primary children as compared to a matched group of non-hearing impaired children. *Psychology in the Schools, 23,* 95–99.

Bracken, B. A., & Fouad, N. (1987). Spanish translation and validation of the Bracken Basic Concept Scale. *School Psychology Review, 16,* 94–102.

Breen, M. J. (1985). Concurrent validity of the Bracken Basic Concept Scale, *Journal of Psychoeducational Assessment, 3,* 37–44.

Center for Science, Mathematics, and Engineering Education. (1996). *National science education standards.* Washington, DC: National Academy Press.

Cummings, J. A., & Nelson, B. R. (1980). Basic concepts in oral directions of group achievement tests. *The Journal of Educational Research, 50,* 159–261.

Flanagan, D. P., Alfonso, V. C., Kaminer, T., & Rader, D. E. (1995). Incidence of basic concepts in the directions of new and recently revised American intelligence tests for preschool children. *School Psychology International, 16,* 345–364.

Frayer, D. A., Frederick, W. C., & Klausmeier, H. J. (1969). *A schema for testing the level of concept mastery.* Working Paper from the Wisconsin Research and Development Center for Cognitive Learning, The University of Wisconsin.

Howell, K. K., & Bracken, B. A. (1992). Clinical utility of the Bracken Basic Concept Scale as a preschool intellectual screener: Comparison with the Stanford-Binet for Black children. *Journal of Clinical Child Psychology, 21,* 255–261.

Kaufman, A. S. (1978). The importance of basic concepts in the individual assessment of preschool children. *Journal of School Psychology, 16,* 208–211.

Lehn, B. (1999). *What is a scientist?* Brookfield, CT: The Millbrook Press.

McIntosh, D. E., Brown, M. L., & Ross, S. L. (1995). Relationship between the Bracken Basic Concept Scale and Differential Ability Scales with an at-risk sample of preschoolers. *Psychological Reports, 76,* 219–224.

McIntosh, D. E., Wayland, S. J., Gridley, B., & Barnes, L. L. B. (1995). Relationship between the Bracken Basic Concept Scale and the Differential Ability Scales with a preschool sample. *Journal of Psychoeducational Assessment, 13,* 39–48.

Novak, J., & Gowin, B. D. (1984). *Learning how to learn.* New York, NY: Cambridge University Press.

Panter, J. E. (2000). Validity of the Bracken Basic Concept Scale—Revised for predicting performance on the Metropolitan Readiness Test—Sixth Edition. *Journal of Psychoeducational Assessment, 18,* 104–110.

Panter, J. E., & Bracken, B. A. (2000). Promoting school readiness. In K. M. Minke & G. G. Bear (Eds.), *Preventing school problems—Promoting school success: Strategies and programs that work* (pp. 101–142). Bethesda, MD: NASP.

Panter, J. E., & Bracken, B. A. (in press). Validity of the Bracken School Readiness Assessment for predicting first grade readiness. *Psychology in the Schools.*

Rhyner, P. M., & Bracken, B. A. (1988). Concurrent validity of the Bracken Basic Concept Scale with language and intelligence measures. *Journal of Communication Disorders, 21,* 479–489.

Rutherford, F. J., & Ahlgren, A. (1989). *Science for all Americans.* New York, NY: American Association for the Advancement of Science.

Schaffer, D. (1999). *Mealworms.* Mankato, MN: Bridgestone Books.

Scholastic. (2007). *Scholastic children's dictionary.* New York, NY: Author.

Stebbins, M. S., & McIntosh, D. E. (1996). Decision-making utility of the Bracken Basic Concept Scale in identifying at-risk preschoolers. *School Psychology International, 17,* 293–303.

Sterner, A. G., & McCallum, R. S. (1988). Relationship of the Gesell Developmental Exam and the Bracken Basic Concept Scale to academic achievement. *Journal of School Psychology, 26,* 297–300.

Taba, H. (1962). *Curriculum development: Theory and practice.* New York, NY: Harcourt, Brace.

VanTassel-Baska, J. (1986). Effective curriculum and instructional models for talented students. *Gifted Child Quarterly, 30,* 164–169.

VanTassel-Baska, J., & Little, C. (Eds.). (2003). *Content-based curriculum for gifted learners.* Waco, TX: Prufrock Press.

Wilson, P. (2004). A preliminary investigation of an early intervention program: Examining the intervention effectiveness of the Bracken Concept Development Program and the Bracken Basic Concept Scale—Revised with Head Start students. *Psychology in the Schools, 41,* 301–311.

# Next Generation Science
# Standards Alignment

| Cluster | Next Generation Science Standards |
|---|---|
| From Molecules to Organisms: Structures and Processes | K-LS1-1. Use observations to describe patterns of what plants and animals (including humans) need to survive. |
| | 3-LS1-1. Develop models to describe that organisms have unique and diverse life cycles but all have in common birth, growth, reproduction, and death. |
| Ecosystems: Interactions, Energy, and Dynamics | 2-LS2-1. Plan and conduct an investigation to determine if plants need sunlight and water to grow. |
| Engineering Design | K-2-ETS1-3. Analyze data from tests of two objects designed to solve the same problem to compare the strengths and weaknesses of how each performs. |